The Timeless Energy of

THE SUN

The Timeless Energy of
THE SUN
for Life and Peace with Nature

PREFACE BY
FEDERICO MAYOR

COMPILED AND WRITTEN BY
MADANJEET SINGH

IN COOPERATION WITH

Tapio Alvesalo (Finland) • William H. Avery (U.S.A.)
Boris Berkovski (UNESCO) • Eric Brus (U.S.A.)
W.W.S. Charters (Australia) • Harijono Djojodihardjo (Indonesia)
Peter E. Glaser (U.S.A.) • A. Goetzberger (Germany)
José Goldemberg (Brazil) • Richard Golob (U.S.A.)
Martin Green (Australia) • J.C. Kapur (India)
René Karottki (Denmark) • Gibson Mandishona (Zimbabwe)
Even Mehlum (Norway) • Stanford R. Ovshinsky (U.S.A.)
François Pharabod (France) • David A. Rezachek (U.S.A.)
Albert Mitjà i Sarvisé (Spain) • Ali Sayigh (England)
Scott Sklar (U.S.A.) • Harry Tabor (Israel)
Mineo Tatsuta (Japan) • Jorge M. Huacuz Villamar (Mexico)
Jiang Xinian (China)

U N E S C O

WORLD SOLAR PROGRAMME 1996-2005

Sierra Club Books - San Francisco

Library of Congress Cataloging-in-Publication Data available upon request.

10 9 8 7 6 5 4 3 2 1

1. (page 2) In sub-Saharan Africa the sun makes its power felt in biomass-related dances, especially at the start of the rainy season and the harvest. The strands of straw attire worn by these masked dancers in Mali represent sun rays, and the ritual has acquired a new dimension since the dancers began performing in front of the newly installed photovoltaic panels.

Picasso's Solar Dove of Peace.

Contents

Preface

*D*eveloping a new and sustainable energy economy is one of the major challenges facing humanity as it prepares to enter the twenty-first century. Industrialization and the increasing consumption of fossil fuels, spurred on by the demands for higher living standards from an exponentially growing population, are polluting our oceans and atmosphere, denuding forests, producing holes in the ozone layer of the Earth's stratosphere that protects us from harmful ultraviolet radiation, and creating a risk of global warming. Damage to the Earth's environment, largely due to affluent lifestyles in the industrialized countries, is compounded by abject poverty in many parts of the world, closely linked as it is to overpopulation and inefficient energy use.

International action to promote renewable energies — of the kind initiated by UNESCO with its World Solar Programme 1996–2005, adopted by the World Solar Summit in Harare, Zimbabwe, in September 1996 — is all the more urgent as more than 2 billion people, one in every three persons worldwide, do not have such an elementary necessity of life as electric power. Many of these men, women, and children live in remote regions or island habitats where solar energy is their only recourse. Photovoltaic and other on-site solar energy installations can change the lifestyle of people in faraway locations who cannot be reached through national electricity distribution grids. Many rural communities are at last beginning to enjoy the benefits of electricity, bringing them into contact with the outside world through telecommunication systems, radio, and television. Solar lights installed in schools and clinics have improved the quality of health care and educational services for children. Other sources of solar energy, such as wind turbines, are similarly benefiting both urban and rural communities.

The spin-off benefits of renewable energies are also of the greatest importance. Small and dispersed solar energy projects in rural areas, including individual photovoltaic plants and small hydro facilities, have a cardinal role to play in halting the increasing rush to the cities by peasants living in the poorest regions. Such projects, by virtue of their local and participatory nature, are also more "democratic," tending to create new cooperative structures that resist the concentration of power in a few authoritarian hands. By promoting sustainable development based on partnership with nature, they protect the environment and favor the emergence of a culture of peace, which is inseparable from democracy and development.

Solar energy technology is perceived by many populations as being in harmony with their cultural traditions. Ancient societies invariably paid homage to the sun as a symbol of truth, justice, and equality; as the fountainhead of wisdom, compassion, and enlightenment; as the healer of physical and spiritual maladies; and above all as the source of fertility, growth, and life's renewal. Sustainable energy, with its peaceful, participatory, and environmentally friendly associations, seems poised to repair the long-standing divorce between science and culture through the traditional use of heliotechnology, the principal energy of the future.

Federico Mayor
Director-General, UNESCO, Paris

---

3. *Photovoltaic power from the Kodari/Tatopani Solar Power Project illuminates this Tatopani Buddhist monastery in Nepal, overlooking the Chinese village of Khasa across the Bagmati River border between the two countries. Solar energy has brought science, education, and culture together, as Buddhist lamas have placed a stone sculpture of the bodhisattva Vajrasattva (center) to protect the PV panels and are also teaching heliotechnology in their monastery's open-air school.*

4. *South Indian sun symbol.*

6

The Energy of the Sun
for Life and Peace with Nature

A charming episode in Mâui-akamai's adventurous life, echoing similar solstice myths in ancient cultures worldwide, tells how the Polynesian demigod harnesses the energy of the sun to help his mother, Hina-a-keahi — "Energy of the Fire" — dry the robe (*kapa*) she had made for him by soaking and pounding the *wauke* tree bark. The days being short, the sun would race across the sky before Hina-a-keahi could dry her son's *kapa*. So she asks him to find a way of making the days longer. Solstice comes from the word *solstare,* or "stopping of the sun." This is the impossible mission with which Mâui is charged by his mother so that she may tap the solar energy not only for herself but for their kinfolk as well. Mâui has no idea how to go about it and so Hina-a-keahi advises him to draw on the traditional experience of his grandmother Mauie, who lives in a cave on the way to the great volcano Haleakalâ — the "House of the Sun."

A mountain is the abode of the sun in many cultures. The volcano of Bojonegoro in Java is venerated because "the sun comes out of it." In Sri Lanka, Saman — the "Energy of the Morning Sun" — is worshipped on Samanolakanda, also known as Adam's Peak. The Chinese Emperor Wu-di of the Han Dynasty (140–87 B.C.) frequented the Cheng mountain at the eastern end of the Shandong Peninsula in order to worship the power of the sun rising from the sea. The mountain deity of ancient Japan, Yama-no-kami, is worshipped as "the solar energy of forests and animals." Pausanias, a Greek traveler in the 2nd century, relates how he saw altars of sun worship on the highest acropolis at Corinth, called Akrocorinth. The Scythian nomads worshipped the energy of fire on their sun-related mounds, called *kurgans.* Machu Picchu in the Andes Mountains in Peru is identified with the sun just as the mountain-like pyramids in pharaonic Egypt and at

5

---

5. A Vietnamese girl uses the passive heat of the sun to dry her cookies, while sunshine provides her with electric power through PV panels installed on the roof of the hut.

6. Woodcut prints of Mâui-akamai's solstice legend by the artist Dietrich Varez. Courtesy: Bishop Museum, Honolulu, Hawaii.

9

Teotihuacan in Mexico are solar symbols. Homage was paid to the sun-mountain in Central Asia, as depicted in the famous victory stele from Susa (2250 B.C.) of the Akkadian king Naram-Sin.

After tediously climbing the steep Haleakalâ mountain, Mâui reaches his grandmother's cave. The cave is the traditional home of the sun worldwide. It is out of a cave that the Japanese sun goddess, Amaterasu, emerges to illuminate the world at the Futamigaura seashore in Japan. The Australian aborigines believe that the sun is a beautiful woman who sleeps in a cave during *wongar*, or "dreaming times," and similar sun-related cave and netherworld legends exist among the Austro-Asian peoples. In the European Celtic tradition, the home of the sun is an immense cave in which the so-called Grand Sourcil, or "Great Eyebrow," goes to rest each evening. The Incas in Peru believe that the sun lives in a cave with

their ancestors, the "Origin People," and lets them come out of it from time to time. The Aztecs' sun god lives in a cave together with the root of a tree.

Entering the cave, Mâui finds his grandmother roasting energy-giving bananas on fire taken from the great volcano Haleakalâ. This is the sun's breakfast, which Mauie cooks every morning as soon as her rooster crows to announce the sunrise. Paying his respects, Mâui asks her how he could slow down Kalâ, the sun. Mauie, who wants to dry her *kapa* as well, agrees to help and, giving him a strong rope, tells her grandson to hide in the roots of the old *wiliwili* tree near her cave and to lasso Kalâ's legs as soon as he appears on the eastern rim of the volcano.

Feeding the sun to rejuvenate its energy at sunrise, as well as at the turn of the seasons, is an ancient belief in all cultures. There is an obvious analogy with muscle power, the primal

8

9

7. *Aborigine village chief Syd Coulthart is a competent technician who is in charge of the 3.5-kilowatt PV system installed in the Ukaka community in the central Australian desert.*

8-9 *The sun is "energized" in many cultures by building wooden towers and then setting them on fire, as seen at Urbes, Vallée de Thann, Alsace, France, during the festival of St. Jean.*

energy source of humankind, which remains as important in most parts the world today as during the Stone Age. Mauie has no doubt that if she stops feeding Kalâ every morning, he would weaken and reduce the solar power so indispensable for the survival of humankind — the sun being the unique energy source that sustains and links all life on the Earth.

Similarly, a French peasant tradition relates how the sun, having been exhausted by his day-long journey, goes to the netherworld after sunset to rest in his mother's house, where she prepares for him a large and revitalizing meal so that he may regain his strength. The Romans, like the Aztecs in Mexico, offered lavish feasts to their sun god, Mithra, during the winter solstice festival — now celebrated as Christmas. Another successor to a pagan festival, Easter, is observed in Bulgaria by eating "sun-bread" decorated with sun symbols and ritually presented to neighbors and relatives before being placed on the graves of ancestors. The tradition is especially strong in Nordic countries, where midsum-

mer solstice rites are still performed with bonfires in order to "energize the endangered sun when it becomes tired and weak" at the turn of seasons, although a Christian message is now assigned to this ancient sun ritual.

The sun/fire energy symbolism of a rooster that announces the sunrise can be found in many cultures. The Yoruba people of Benin, in Africa, believe that roasting a rooster on the fire enhances their sun-related powers of fertility, growth, and life renewal. An ancient Byzantine ritual is still observed in parts of Central Asia, where people sacrifice a rooster and energize the sun by lighting fires on hilltops on the day — July 20th — that the Prophet Elijah emerged from the dire seclusion of his cave to ascend to heaven in the solar chariot pulled by the fiery energy of winged horses. Prior to his ascension, Elijah (a paleo-Hebrew personage revered by Judaism, Christianity, and Islam alike) washes his skin coat, dries it in the sun, and gives it to his disciple Elisha — much as Mâui's mother washes his *kapa* and needs solar energy to dry it.

The renewable aspect of solar energy and fire is symbolized in Egyptian tradition by the

10. In many cultures, solar energy is symbolized by mythical birds, such as these painted on the 17th-century Bukhara gateway, Uzbekistan.

11. As in an ancient Byzantine ritual in Central Asia, the Yoruba people of Benin in Africa also believe that the sun is energized by sacrificing a chicken, as represented by this wooden piece.
Courtesy: Stanley Collection of African Art, University of Iowa Museum of Art, Iowa City, U.S.A.

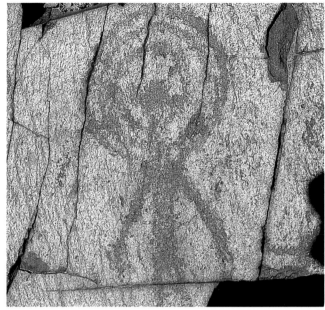

heron, *bennu*, the fabulous bird of classical antiquity. Later identified with the phoenix, the golden sun bird with brilliant scarlet plumage said to live for 500 years, makes a nest of aromatic boughs and spices, sets it on fire, and is consumed in the flames when its end approaches. The Egyptian legend has it that a new phoenix then springs miraculously out of the pyre and, carrying the embalmed ashes of its father in a golden egg of myrrh, flies to Heliopolis to deposit them on the altar of the Temple of the Sun. The image of the phoenix also appears on the coinage of the late Roman Empire as the symbol of the Eternal City of Rome; and in another remote corner of the world, solar energy is represented in a beautiful Aztec sculpture as a sun deity in the guise of an eagle.

As the rooster crows, Mâui hurries out and hides behind the wiliwili tree until he sees the sun peeping over the edge of the mountain to start racing across the sky on his daily errands. Waiting until Kalâ leaves the horizon and comes close over his head, Mâui leaps out of his hiding place and, flexing his renowned muscle power, throws his lasso over the sun. But Kalâ is not so easily captured; his solar energy is the equal of Mâui's strength, and the struggle continues for a long time as the sun tries to escape.

12. The sun in human form represents muscle power in this Bronze Age rock drawing at Saimaly Tash, Kyrgyzstan.

13. A mountain is the abode of the sun in many cultures, and its power is represented by this victory stele of the Akkadian king Naram-Sin from Susa, c. 2250 B.C.
Courtesy: Louvre Museum, Paris.

14. *The river goddess Daphne prays to Earth (her father) for help and turns into a laurel of peace when pursued by the Greek sun god Apollo. Courtesy: Borghese Gallery, Rome.*

15-16 *The intimate relationship between the sun and the tree (biomass) is shown by these two 8th-century B.C. Phoenician ivory panels. Courtesy: British Museum, London.*

17. *Solar power is traditionally celebrated in Bulgaria by eating "sun bread," decorated with sun symbols and ritually placed on the graves of ancestors.*

18. *A number of remote villages in the Indian state of Rajasthan, including Sam, Sadrau, Devi Kot, Unda, Kathoda, and Ridwa, have installed PV systems and the panels are often carried on camel-back.*

Had Mâui-akamai not received a traditional education in the sacred land of Kuaihelani, he would never have accomplished his Herculean mission of capturing the sun. He was carried to Kuaihelani by wind power soon after his birth, when his mother, Hina-a-keahi, wrapped him in a lock of her long hair and set him adrift on the sea. There he learned from the gods Kâne and Kanaloa that the sun is the source of energy-giving biomass in the form of *ohe*, or bamboo, which could be used for making fishing poles, carrying water, or building houses, and whose shoot could be eaten. On his return home, his mother's cousins, the

17

alae birds, told him the secret that thermal energy could be obtained by rubbing two dry wooden surfaces together to produce fire. He then discovered how the steam energy produced by throwing heated lava rocks in the river Wailuku is even more powerful than the energy produced when its tranquil waters are stirred by Kuna, the giant eel.

And once, when the sky descended so close to the Earth that there was hardly any space left for people to move about or even breathe, Mâui came to their rescue and used his formidable muscle power to protect the environment by pushing the sky high above the great volcanic mountain Haleakalâ.

18

Red Energy of Muscle Power

Like Mâui-akamai, the hunting and food-gathering people of the Stone Age traditionally relied on their own muscle power, for they had no need to exploit the involuntary or bonded labor of another person or group — whether slave, serf, indentured servant, or otherwise. Those hunter societies that have survived to modern times — such as the Amerindians and the Australian aborigines — seem to have never practiced slavery, which made its appearance only when communities reached the pastoral and later agricultural stages of development.

With the passage of time, the human communities most favored by geography and climate began to make the transition from hunting to a more settled way of life dependent on animal husbandry and agriculture. To bring about this transition from the Paleolithic to the Neolithic (or New Stone) Age (c. 8500 B.C.), more energy was needed, and its availability was determined by the basic tools at their disposal and the materials with which tools were made. Assisted by the additional energy provided by plants — the "biological solar collectors" — farming developed to support more human settlements that grew and spread out, governed and restricted only by the local potential to create energy. Domestication of plants and animals eventually led to improved methods of cultivation and stock breeding and the proliferation of the crafts, which in turn

eventually produced a surplus of energy and thereby freed some of the population to work as artisans, craftsmen, and service workers.

As human settlements increased in size by virtue of technological advances in irrigation and cultivation, the energy needed for improving the circulation of goods and people became ever more acute. Neolithic man, having achieved the domestication of animals, used them for transportation as well as for food and hides. Then came the use of draft animals in combination with a sled equipped with runners for carrying heavier loads. However, a move away from self-reliance to dependence on the muscle power of others appears to have begun with the invention of the two-wheeled wooden cart — used first in the Tigris-Euphrates Valley around 3500 B.C. — constructed first with solid wheels and later with hubs, spokes, and rims. The wheel appears to have been modeled on the circular image of the ever-rolling sun disk in the sky shown on petroglyphs found at several Neolithic and Early Bronze Age sites. In a variety of wheel-like rock drawings, the circle is either plain, crossed in the middle, or actually spoked. To be used efficiently, wheels require roads, and the Mesopotamians were probably the first serious road builders who developed a trade route from the Babylonian Empire west and southwest to Egypt — an art later developed most highly by the ancient Romans.

Another important energy-related discov-

---

19. Muscle power is represented by this Roman-period Celtic wheel god in bronze, armed with his solar wheel, thunderbolt, and lightning flashes. Courtesy: Musée des Antiquités Nationales, Saint-Germain-en-Laye, France.

20. Ancient Romans symbolized their power and glory by this 2nd-century B.C. massive marble sun disk, now called Bocca della Verità. Courtesy: Church of Santa Maria in Cosmedin, Rome.

21

22

23

ery was the use of fire to produce metallic oxides — which led to the transition from the Stone Age to the Metal Age. Although iron, as a scarce and precious metal, had found limited use in the Near East as early as 3000 B.C., by around 1000 B.C. knowledge about iron metallurgy had become widespread. This is reflected in the sun-related Greek legend in which the god Prometheus steals fire from the blacksmith Hephaestus, a myth that is still celebrated annually during the Promethean Festival in Greece. And in a similar African tradition in Mali, Nommo, the companion of the supreme being Amma, "steals a piece of the sun" in order to provide fire for the Dogon blacksmiths. In Europe, the power of solar energy during the Metal Age is well represented by the Trundholm Chariot, now in Copenhagen's National Museum, on which a sun disk is being hauled by a horse.

The horse, harnessed with a rigid collar, horseshoes, and stirrups, became the instrument of this power revolution as the search for

additional energy sources intensified. The energy gap was filled by more efficient harnessing of the muscle power of the horse, which was transformed from an ancillary beast of burden — like oxen, mules, camels, and donkeys — into a highly versatile source of power in peace and war. It was on horseback that the nomadic Cossacks, Scythians, Sarmatians, Aryans, Khazars, Huns, Tartars, and Mongols of Genghis Khan conquered vast regions of Central Asia and the European steppes — a feat also performed by the powerful steeds of Alexander the Great when the Macedonian warrior confronted the Kshatrian warriors in India. The chariot of the sun in which the Greek sun god, Helios, rides is pulled by the muscle power of the horse, and according to the *Rig Veda*, the chariot of the sun deity, Sûrya, in India is pulled by "the horse of the yonder sun who rises from the water like a spark of life at daybreak, and then dies at sunset before rising again the next morning."

With the steady increase in population,

21. *An outstretched hand represents sunrise in a gesture of blessing and peace in most cultures, as shown by this Hand of Fatima.* Courtesy: Islamic Museum, Aqsa Mosque, Jerusalem.

22. *Hands and feet represented the primary working tools of prehistoric Neanderthal man's muscle power.*

23. *This massive 12th-century stone sculpture of a foot is*

engraved with symbols representing the energy of solar muscle power. Courtesy: Mrinalini Sarabhai Collection, Ahmedabad, India.

24. *Muscle power is still the primary source of energy in most developing countries, even though in some places it is now supplemented with new forms of solar energy — as in the Indonesian village of Sukatani, West Java, where 230 homes have been fitted with 85 individual 40-watt solar home systems (SHS), 20 streetlights have been installed, and a television set added to the community center.*

however, there came a time when animal muscle power could no longer satisfy the growing demand for additional energy. Since human muscle power was another available option in the absence of appropriate technology, different types of slavery emerged depending upon the particular stage of development in each society. Among the first to practice slavery were the "cradles of civilization" — the Old Kingdom of Egypt (c. 2686–c. 2160 B.C.) and the Babylonians, at least from the third dynasty of Ur (c. 2100 B.C.). The Greeks practiced so-called praedial, or plantation, slavery, mostly based on the recruitment of barbarians. The immensely wealthy Roman upper classes, during the 2nd and 1st centuries B.C., followed the Greek practice as greater muscle power was indispensable on their huge estates, or *latifundia*. In Spain, too, about 40,000 slaves are said to have been employed at this time in the silver mines of New Carthage.

The dire need for energy was among the reasons why slavery flourished in many parts of Asia in one form or another. In India it was sanctioned by religion in the Laws of Manu (variously dated between the 2nd century B.C. and the 2nd century A.D.), in much the same way as apartheid in South Africa became the instrument of political oppression. The traditional Japanese organization, the manor, or *shôen*, had at least three types of slaves under the lord's absolute control. Servitude was well established in Arabia when the Prophet Muhammad began to preach a new religion in the first years of the 7th century, and his attitude — especially toward women — was similar to that of the Greeks, Romans, and the Christian churches: he considered slavery to be a necessary evil, while teaching that slaves should be treated with humanity. Serfdom in China was widespread during the absolute monarchy of the Ch'in Dynasty (221–206 B.C.), when Emperor

25. The European colonialists profited from the muscle power of their subjects' indentured labor and from the wind energy that wafted their galleons. Oil on canvas of Fort William in Calcutta by George Lambert and Samuel Scott, c. 1732. Courtesy: British-India Office Library, London.

26

Ch'in Shih Huang-ti completed the gigantic monument to slavery, the Great Wall of China. The tremendous amount of human muscle power expended on what turned out to be an unnecessary defensive barrier was as inhuman as the expenditure of the energy of the tens of thousands of slaves whom the Egyptian pharaohs (c. 2686–c. 2160 B.C.) employed in the construction of the great pyramids.

The tremendous energy requirements of the emerging nation-states of Portugal, France, Spain, the Low Countries, and Great Britain virtually institutionalized slavery soon after they discovered the sea routes around Africa's southern coast (1488) and sailed to America (1492). Chinese and Indian indentured labor was exploited in the lands colonized, moving the Chinese men great distances in Asia and importing slaves from Africa for the benefit of the New World plantation economies. Humans virtually became beasts of burden and, profiting from their surplus muscle power, the colonialists accumulated enormous wealth as they used wind energy to waft their galleons worldwide from the 16th to the 18th century. Great Britain benefited most from this unprecedented prosperity, and from there it spread to other parts of Europe and America.

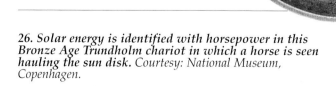

27

26. Solar energy is identified with horsepower in this Bronze Age Trundholm chariot in which a horse is seen hauling the sun disk. *Courtesy: National Museum, Copenhagen.*

27. The muscle power of the horse created the myth of the Greek sun god Helios riding in his celestial chariot, as seen in this 3rd-century B.C. silver plate. *Courtesy: Hermitage Museum, St. Petersburg.*

Black Energy of Fossil Fuels

With no large slave labor force to draw upon in their own countries during the colonial period, Europeans experienced a labor shortage that stimulated a search for alternative sources of energy and the introduction of labor-saving devices. This process of change from an agrarian, handicraft economy to one based on industry and machine manufacture started with the use of coal and later became centered on petroleum and natural gas — which are also forms of solar energy, since their power was derived over millions of years from sunlight in living organisms. These nonrenewable black energies eventually laid the foundation for the Industrial Revolution.

The Chinese are said to be the earliest to have extracted coal to supplement natural gas and their dwindling supply of wood, and the Romans also used it as a fuel to some extent. But it was not until the 13th century that coal become the object of European commerce, together with the wood that continued to be the primary fuel for heating, cooking, and charcoal production. With supplies of wood dwindling, coke from coal began to replace charcoal in 1709. The new coke industry, together with the production of coal gas for illumination, led to a rapid expansion of the coal-mining industry, which produced it in the forms of lignite, anthracite, bituminous, and sub-bituminous coal. Today about 1.5 trillion tons of coal is recovered worldwide.

Coal remained the world's main source of energy from the late 19th century until the early 1960s, when the pace of technological developments required much more energy than coal could provide. Petroleum, known to the Babylonians though not extensively used, first came into prominence in Europe when an oil well drilled in 1640 at Modena, Italy, was used to provide fuel for street lighting. More oil wells were developed in Romania in 1650. Then, following the discovery of a major oil field in western Pennsylvania in 1859, the availability of this inexpensive and more efficient energy source increased the demand for liquid fuels, principally for use in kerosene lamps. This was followed in the latter half of the 20th century by a huge demand for oil in the automobile industry, electric-power generation, and in the manufacture of synthetic chemicals and other oil-based products.

Between 1950 and 1984, world energy demand increased four-fold — the sharpest rise occurring between 1960 and 1970. In this period the demand for coal as a source of energy continued to decline as coal-fired ships began disappearing from the oceans. Oil-fired electricity generation began to compete with coal-fired plants, and with the invention of a practical internal-combustion engine, the consumption of oil increased rapidly, doubling every 15 years. Today about 80 percent of oil is consumed in North America, western Europe, the former

8

28. The relationship between energy generation and global warming associated with hydrocarbon fuel combustion, which produces the "greenhouse effect," is potentially catastrophic. The causes and consequences of climate change are global and require a global response, just as arms control treaties ought to limit and reduce nuclear arsenals.

29. A Chinese sun symbol with a black raven. Detail from 2nd-century B.C. T-shaped painting found in Tomb 1, Ma-wang-dui village, Changsha city, Hunan, China.

Soviet Union, and Japan, while five Middle Eastern countries — Iran, Iraq, Kuwait, Saudi Arabia, and the United Arab Emirates — have become the source of nearly two-thirds of the world's oil production, with Saudi Arabia alone accounting for more than 25 percent of this.

The ever-increasing exploitation and limited reserves of fossil fuels have now convinced even the most enthusiastic supporters of coal and oil that the remaining life span of both is relatively short. Although new reserves are discovered from time to time, the era of fossil fuels will be no more than a momentary flicker in the world's 5-billion-year history. At current rates of production, the world's proven oil reserves will be exhausted in about 50 years, although coal reserves may last for about 230 years. Jules Verne had predicted as much in 1877, in a book entitled *Les Indes noires*. Wilhelm Ostwald, a Nobel Prize winner in chemistry, took up the problem in 1909, pointing out in his book *Energetische Grundlagen der Kulturwissenschaft* that the "capitalized stockpiles in the form of fossil fuels are like an unexpected inheritance, persuading the inheritor to lose sight of the principles of a lasting economy and to live for the day"... while not realizing that "even thrifty consumption merely postpones, but does not prevent its exhaustion."

The energy crisis caused by the decision of the Oil Producing and Exporting Countries (OPEC) to raise the price of oil in 1973 made the world realize for the first time that it is living in a "Petrolithic Age," ruled by the unwritten laws of its oil prices and pollutants. Revolution in Iran and the outbreak of war between Iraq and Iran caused a further price increase in 1979, making oil 17 times more expensive than in 1972. Iraq's surprise invasion of Kuwait in August 1990 served as another chilling reminder of the economic and political dangers of the world's continuing dependence on oil. Already the signs are that the "hidden expenditure" of some 60 billion dollars spent on protecting the strategically important oil reserves of Kuwait, Saudi Arabia, and other Gulf states might turn out to be as fruitless as the money squandered on the Great Wall of China by Emperor Ch'in Shih Huang-ti, whose Tartar enemies had already bypassed this defensive barrier before it was even completed.

As well as witnessing the effects on American and allied soldiers of the "Gulf War (oil) syndrome" produced by poisonous gases, millions of television viewers all over the world became aware of the enormous environmental destruction caused by the war. Each day relentless flames devoured some 5 million barrels of oil, generating more than half a million tons of airborne pollutants, including sulphur dioxide, the key component of acid rain. Billowing almost 2 miles (3 kilometers) high, the sooty pall rode the wind as black rain, falling not only in Kuwait but far beyond in Saudi Arabia, Iran,

India, and Pakistan. Numerous accidental oil spills, such as the *Exxon Valdez* wreck in Prince William Sound off the coast of Alaska in 1989, have polluted the seas with horrifying regularity, causing enormous economic and ecological damage through the destruction of fish, birds, and other forms of life on which the world's aquatic system depends.

In the polluted atmosphere of overpopulated cities, more and more people are suffering from ill health, and historic buildings that have survived for hundreds of years — such as the Coliseum in Rome, the Parthenon in Athens, and the Taj Mahal in Agra — are now suffering from the effects of increasing levels of air pollutants and acid rain, which are heavily corroding the limestone and marble with which these beautiful monuments were built. People everywhere are alarmed at increasing water contamination, air pollution, the burning of vegetation, the poisoning and erosion of soils, the extermination of biological species, and the depletion of the ozone layer that protects our environment by absorbing harmful ultraviolet radiation.

After years of debate, no one can seriously dispute the existence of global warming — created by the burning of fossil fuels in power sta-

tions, automobiles, and homes — which releases carbon dioxide into the atmosphere, trapping more of the sun's warmth than the planet would otherwise retain. Testifying to the U.S. Senate in 1988, James Hansen, professor of geological sciences at Columbia University, stated that our planet has been cooling slowly and irregularly — the coldest time being the Little Ice Age of 1600 to 1850. Since the Industrial Revolution began, the long-term cooling trend has reversed, and Earth, on average, has warmed by about 1 degree Fahrenheit (half a degree Celsius) since 1850, with the most rapid warming occurring since the middle 1960s. The greenhouse effects of "global warming" are far-reaching, as some parts of the world are likely to become colder and others hotter; monsoon and hurricane paths may shift, and storms may become more violent, while the sea level may rise to flood large areas of the world's coastline.

Equally dangerous is the radioactive contamination caused by the storage of nuclear waste. Plans to build a central storage site in the southeastern corner of New Mexico by the U.S. Department of Energy is mired in controversy. The subterranean dump for low-level energy waste to be stored almost half a mile

---

30. Some 60 billion dollars spent on the Gulf War is just one example of the unaccounted "invisible cost" of obtaining fossil fuels, besides causing such health hazards as "Gulf War (oil) syndrome," from which American and Allied soldiers are still suffering.

31. Global warming is threatening the Earth's environment, which had remained stable for the past 10,000 years. The enormous ecological disaster wrought by the widespread effects of the sooty pollutants of acid rain, including sulphur dioxide, are destroying the environment, and the economic and human costs may devastate continents.

(800 meters) underground in the salt rocks has been decried by environmentalists as a potential "Swiss cheese" because it is feared that at some point in the future, people drilling for oil or water might puncture the dangerous radioactive brine pool beneath the salt formation. Only after 240,000 years would this kind of waste lose its harmful radioactivity.

On the tenth anniversary of the Chernobyl accident that spewed a deadly cloud of radiation across Russia and much of Europe, the French government ominously handed out iodine pills to 400,000 people living near nuclear power plants as a safety precaution; an additional 2 million pills are stockpiled in Paris warehouses in anticipation of a major incident. A study in the *British Medical Journal* recently reported that children playing near a French nuclear waste processing center in Normandy face a risk of developing leukemia. At its La Hague plant west of Cherbourg, the state-owned company Cogema processes waste from more than 50 nuclear power plants across France and from the German and the Japanese nuclear industries.

On the other side of the world, the South Koreans have protested in large public demonstrations against the deal under which North Korea, in exchange for tens of millions of dollars, has agreed to accept thousands of barrels of nuclear waste from Taiwan for dumping at a site about 40 miles (65 kilometers) across the border between the divided country. A similar proposal was dropped when the residents of the

32. More than 2,500 scientists of the United Nations Intergovernmental Panel on Climatic Change concluded in 1995 that unless the current rate of combustion of carbon-based fuels — coal, gas, and oil — is reduced, worldwide temperatures would rise an average of more than 3° Fahrenheit (2° Celsius), creating a crisis larger than any other in recorded history.

33. The expression of desolation on the faces of these Indian coal miners in Bihar is typical of the millions of such workers worldwide. Since coal became the object of European commerce in the 13th century, their living conditions have remained deplorable.

Marshall Islands objected to the disposal of Japanese nuclear waste in their country, and more recently, in March 1997, a fire and explosion at the spent fuel reprocessing facility at Tokai raised a public outcry against the "deadly fire of the atom." Sensing the universal public outrage, the Swedish parliament recently decided unanimously to phase out nuclear power and make the country nuclear-free by 2010.

Continued increases in global energy demand and consumption, a growing concern about environmental problems, and the development of new industries are three of the principal factors influencing investment in advanced renewable energy systems and technologies. Such technologies already have the potential to far exceed current global energy requirements, while the whole cycle of production, storage, and transmission of renewable energy has only about 1 percent of the overall environmental impact of fossil fuel–based energy production. Since it is an area of highly innovative technology, specialized skills are essential, so the sector has the additional potential to generate jobs at two or three times the rate expected for conventional industries of comparable turnover.

Although advanced energy technologies — including solar energy, wind energy, hydrogen technology, and various methods of energy storage — currently account for only a minor proportion — around 0.1 percent — of global energy production, they may already be significant in some areas on a local scale. In many remote locations, for example, new technologies are currently providing commercially viable and competitive alternatives or supplements to traditional forms of energy production. Indeed, in certain areas, advanced energy technologies already generate 3–4 percent of the total national energy output. Global wind energy production capacity amounts to the equivalent of 3,000 megawatts, while solar electric power generation and solar heating correspond to 500 megawatts and 20,000 megawatts, respectively.

Careful evalua-

tion of the limited nonrenewable fossil reserves of coal, oil, and natural gas and the dangers of radioactive contamination by nuclear plants have persuaded even some of the oil companies and the OPEC nations to invest more in research and development related to sustainable solar energy. British Petroleum PLC, for example, has recently invested $20 million in what it says will be one of the largest and most technologically advanced solar cell manufacturing plants in the world, to be situated in Fairfield, California. They have come to realize that there is no conflict of interest with their main form of commerce, especially in the isolated regions inhabited by one in every three individuals in the world, where centralized grid installations are impossible in any case. Even in highly populated urban areas, the grid-connected renewable and the nonrenewable energy sectors are mutually compatible.

False alarms that curbs on "global warming" might harm industry have been recently rejected by more than 2,000 economists, led by Nobel laureates Kenneth J. Arrow and Robert M. Solow, who stated that investing in the research, development, and marketing of environmentally friendly, energy-efficient technologies will, on the contrary, open up new areas of economic and commercial opportunities. Their view is perfectly in tune with the belief in most cultures that identifies the traditional energy of the sun with material wealth such as gold. It is symbolized, for example, by a Chinese "piggy bank" in bronze dating from 206 B.C. with a gilded solar deity riding a golden horse surrounded by four muscular oxen — found at Jinning in the Shizhai mountains of the Yunnan Province. The usurpation by black petroleum of the expression "liquid gold" and the false coinage "golden coal" are desecrations of the traditional homage paid to the "sun metal" and an affront to the protection of the environment. Only the sustainable energy of the sun can claim to be called "golden" and possibly succeed in launching a new Golden Age — synonymous with the Green Age — the final stage in the Stone-Bronze-Iron-Age sequence.

34. In our interdependent "global village," protection of the environment must be a cooperative effort. What individuals and nations do affects the economic well-being and even the survival of others. Global warming threatens not only to flood coastal lowlands and tropical islands, and destroy watersheds, but may subject forests and croplands to the ravages of disease, pests, and fires. There is no way out but to halt the misuse of fossil fuels — as suggested by the "Black Hand" designed by Madanjeet Singh.

Green Energy of Biomass

The totem of the subtribe Tau of the Kenyahs in Indonesia is a tree called *tau* (the sun), against which a liana of the lurek plant is rubbed in order to produce fire — recalling the method first used some 1.5 million years ago by the peoples of the tropical and subtropical Old World. In several African countries the ancient ritual in which "male and female" firewood are rubbed against each other to ignite fire is still observed. The practice in Japan of felling a sacred tree on a mountainside, dragging the log down to the village, and erecting it as a holy pillar, *mihashira*, recalls the ancient, biomass-related fertility festival of Cybele — the Greco-Roman mountain-mother — at which a pine tree was felled and brought to her shrine and worshipped along with her lover, Attis. In Serbia and Russia and among the Slavs living near the Elbe River, the sun's energy is identified with old trees, and the Aztecs believe that it is from the root of a tree that their sun gods emerge. In an Indian biomass energy ritual, colorful dolls — identified with people from all walks of life

— are strung around old trees. The famous Buddhist icon of Maya Devi is the nativity of the Buddha in which his mother leans on a tree — a 2,000-year-old tradition still echoed in Dohda ritual in many parts of India and performed in the belief that the harvesting season will not start unless a virgin kicks a tree to make it blossom at sunrise.

Bio means life, so *bio-energy* is energy from living things. The term *biomass* refers to the material from which we get bio-energy. Biomass is produced by the sun's radiation from the photosphere — its luminous surface some 93 million miles (149 million kilometers) away from the Earth — which converts hydrogen into helium at extremely high temperature and pressure. This energy, captured in the form of sunshine by green leaves, is converted into the chemical energy of plant-stuff. The ultimate source of virtually all organic feedstocks is the conversion of sunlight, water, and carbon dioxide into carbohydrates and oxygen by plants and photosynthetic algae.

A feedstock's chemical energy is released directly through combustion, or the

35

37

---

35. Biomass is being consumed at a rate faster than it is being produced worldwide, and even the lush greenery of the Malaysian island of Piamenggil in the South China Sea was in danger of severe denudation until housewives began using biogas for fuel. The use of electrical appliances powered by a 10-kilowatt photovoltaic system installed about five years ago also reduced the excessive exploitation of biomass.

36. The sun is the abode of the Vedic deity Vishnu, the preserver of life, and his consort, Lakshmi, the goddess of fortune. Indian miniature painting (1725). Courtesy: Bharat Kala Museum, Varanasi, India.

37. The harvest season is a joyous occasion worldwide, and the biomass is identified with sun rays as represented by the skirts of straw worn by Polynesian women dancers.

feedstock can be converted into other fuels for later use. Thus, harnessing biomass energy is in effect a way of exploiting "nature's solar collectors" — living plants that use photosynthesis to turn the energy of sunlight into carbohydrate molecules. Humans and animals use that energy whenever they eat vegetable matter.

Plants use sunlight for growth in a process called photosynthesis.

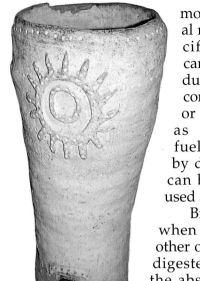

38

In this process, the plant combines carbon dioxide from the air and water from the ground to produce oxygen and energy. Energy is stored in the plant in the form of carbohydrates. This energy is released when the plant is burned or converted into a fuel. When animal waste is converted into gaseous fuels, the manure is broken down by bacteria through an anaerobic (oxygen-free) process. The resulting gas is burned to generate electricity or for local heating. Biomass can also be treated with chemicals, converted by microorganisms, or subjected to high pressures and temperatures to produce liquids, gases, and solids that offer the possibility of replacing petroleum-based fuels.

Biomass consists of growing plants or agricultural waste such as trees, grasses, straw, stalks, husks, cobs, dried dung, manure, ocean plants and even garbage. Until about 1700, these various sources provided most of the energy requirements of the world, and they are still essential for heating and cooking in many countries where other fuels are not available. Wood is still the primary energy source for more than half the world's population. It is the energy source for heating homes and cooking food, while woody material is now also increasingly used as fuel for electric power plants in both the developing and industrialized countries. Farm products such as corn, sugarcane

molasses, agricultural residues, and specific energy crops can be used to produce heat or can be converted to liquids or gases to be used as transportation fuels. Gases formed by decaying biomass can be collected and used as fuel.

Biogas is produced when animal waste or other organic material is digested by bacteria in the absence of oxygen. This can happen where rubbish has been buried underground, just as in the digestive system of humans and other mammals bacteria break down food in the intestine. To produce biogas for energy use, animal manure and/or human sewage, and sometimes food wastes, are mixed with water, stirred, and warmed inside an airtight container called a digester. Biogas is a mixture of gases — methane, carbon dioxide, hydrogen sulphide — plus water vapor — but it is only the methane gas that burns. One ton of food waste produces about 3,000 cubic feet (85 cubic meters) of biogas, which is approximately 60 percent methane; the rest is mostly carbon dioxide. Biogas is a high-quality fuel that is excellent for combined heat and power generating (CHP) plants. It is a CO_2-neutral fuel that can be used as a substitute for fossil fuel consumption. Furthermore, digestion of manure and organic waste causes lower emissions of methane than ordinary open tanks and landfills. Biogas systems are highly suitable for processing liquid manure and industrial waste, which can afterwards be utilized as fertilizer in agriculture. The utilization of farm manure for biogas production contributes to the reduction of nitrate pollution. The organically bound nitrogen is mineralized and becomes available to plants. This leads to improved utilization and less leaching out.

The process is slow and complex and requires certain environmental conditions —

3.

biomass. They are the ones who inaugurated the chipko, or "tree-hugging" movement some 250 years ago, as villagers valiantly resisted the felling of trees by the Jaipur maharaja's soldiers, who brutally hacked to death 363 Bishnois, along with their brave woman leader, Amrita Devi.

the higher the temperature, for instance, the faster the digestion and biogas production. Below 50°F (10°C), biogas production occurs very slowly if at all, while above 150°F (65°C) very few strains of bacteria can survive. The more time allowed for the process, the more complete the digestion and the more biogas will be produced. In addition to the gases produced, the fermentation of organic materials reduces them to a slurry with a high concentration of nutrients, making them especially effective and valuable as fertilizers.

Today in Denmark, more than 500,000 tons of liquid farm manure together with 140,000 tons of industrial waste are digested each year to produce 800 million cubic feet (23.3 million cubic meters) of biogas. The annual electrical energy obtained from this biogas is some 15,000 megawatt-hours. Cooperatives formed by local farmers and individual farmers are already playing an important role in the further dissemination of biogas technology. There are 2,000 large pig-farms that could integrate biogas production into their manure treatment systems and at the same time produce heat and power. In recent years, gas production at several plants has increased two to three times due to supplementing the liquid manure with other forms of organic waste — for example, from the fish-processing industry, dairies, slaughterhouses, households, or sludge from sewage treatment plants.

Two types of biogas systems are presently in use on Danish farms: centralized, cooperatively owned plants and single-farm plants. In the centralized plants, manure from several farms within a radius of around 7 miles (11 kilometers) is collected by truck and sent to digesters with capacities ranging from 20,000 to 250,000 cubic feet (600 to 7,600 cubic meters).

40. Plastic material is extensively used worldwide to preserve passive heat in greenhouses. The picture shows the interior of a greenhouse at the Institute of Agricultural Research in Nicosia, Cyprus.

41

The manure is digested, stored, and then returned to the individual farms for fertilizer. The biogas is often piped to a nearby town to be used for a CHP plant. There are now about 15 such biogas cooperatives, each with 30 to 40 farmers as members. Some of these are now also looking into energy crops from agriculture, such as rapeseed oil and alcohol. In the single-farm plants, the manure is digested on site and the gas used for a CHP plant, which provides heat for the stables and the household. Any excess power is sold off to the utilities.

The growth of large cities was primarily responsible for the development of refuse incinerators. The burning of refuse provided an efficient means of reducing its bulk to a readily transportable and often usable ash. Municipal refuse contains a great variety of combustible materials that require special treatment before burning, and air pollution agencies in many places now regulate the use of such incinerators. Bulky refuse may be reprocessed by size reduction through such equipment as shredders, hammer mills, and impact mills. Hazardous and obnoxious waste, such as highly volatile dusts and flammable liquids, are usually disposed of in specially designed plants. The use of incinerators led to the development of digesters as the heat from incinerator furnaces could be used to generate steam for heating, power generation, and for use in industrial processes.

Europeans have been using digesters to produce biogas from domestic and farm waste and converting it into heat and power for about a century. For example, a combined heat and power project in southeast London converts 420,000 tons of municipal solid waste into 30 megawatts of electricity annually — enough to power about 50,000 homes. This highly efficient

41. The flower zone of Kagoshima is located at the southern tip of the Satsuma Peninsula in Japan, where a large variety of flowers are grown on the seashore.

42. (following pages) Vegetable Garden and the Moulin de Blute- Fin on Montmartre. Oil on canvas by Vincent Van Gogh (Paris, 1887). Courtesy: Van Gogh Museum, Amsterdam.

34

43

44

45

waste-to-energy plant was commissioned in early 1994 and uses state-of-the-art incineration technology. Waste is deposited from trucks in an enclosed store and is transferred to two incineration streams by overhead cranes bringing 29 tons per hour to boilers. Combustion air is drawn from above the refuse store, thus preventing odors from escaping the plant building. High-pressure steam drives a single turbo-alternator, and the gases are cleaned by a lime scrubber, bag filters, and activated carbon before being released into the atmosphere through a 320-foot-high (100-meter-high) chimney. Acids are neutralized, dust trapped, and dioxins and heavy metals are removed in this process. Exhaust steam is cooled by air-cooled condensers, and the resulting hot water is dis-

tributed to some 7,500 houses and schools as part of a district heating scheme.

Metals, glass, and paper are recycled while organic wastes are composted and returned to the soil or to digesters to produce methane gas. In western Germany about a quarter of garbage landfills produce power from biomass energy, while in Great Britain 54 such biomass plants capture about 10 percent of the amount of methane generated and released into the atmosphere each year. Japan uses 85 percent of its waste, and it is estimated that livestock waste could produce enough biogas to reduce the country's oil consumption by about one-quarter of the total amount of oil used annually in Japanese agriculture. Switzerland reuses 80 percent of its waste, and France about 50 per-

43. *The conical straw (biomass) hat worn by these rice planters in Japan is identified with the sun and used by peasants in many parts of Asia; it recalls the sun-hat called petasos worn by the Greek deity Hermes.*

44. *A huge log of wood, representing the sun, is carried to Shicho Kohoka shrine at Ise during a Japanese biomass festival called Okibiki. Tree worship recalls the Greco-*

Roman festival of Cybele, during which a pine tree was felled and worshipped annually at Cybele's shrine.

45. *These rice cultivators in south India use an SPV pumping system to water their rice fields — an agricultural program in which pumps with array ratings from 200 to 2,250 watts are available.*

cent. A South African company uses biogas from municipal dumps to produce chemicals. The city of Honolulu's "garbage to energy" plant burns the refuse-derived fuel to provide electricity that is sold to the Hawaiian Electric Company. In 1993, the plant produced approximately 6 percent of Oahu's electricity. Recovering all such waste materials from agriculture and forestry could one day supply enough energy to meet 7.5 percent of current world energy needs, estimates David Hall, a biologist at the University of London's King's College.

Recent advances in combustion engineering, biotechnology, and silviculture are making it economical to turn a variety of plant forms into usable liquid or gaseous fuels. The latest technology is now used in Sweden, where a new heating plant in the southern town of Värnamo started operating in the winter of 1994. It generates 6 megawatts of power and 9 megawatts of heat for the town's district heating system by gasifying wood and burning it in a jet engine. In all, more than 80 percent of the energy content in the wood is used for heating buildings or powering lights and motors in the town, while emitting no sulphur and only a limited amount of carbon dioxide — which is absorbed by the new trees planted to replace the ones that are cut and burned. Theoretical estimates suggest that suitable techniques would allow biomass to provide about half the world's liquid and gaseous fuels and one-fifth of its electricity.

Finland's industry and foreign trade is largely dependent on

46

its extensive forests, and its entire forestry sector is based on sustainable development. It is among the few countries where forests are protected by strict rules whereby more trees are planted than felled. Environmental considerations also play a major role in the manufacture of paper and board — the most important national products. The Sustainable Paper Technology Program (SPTP), which aims to simplify manufacturing processes for pulp and paper, has achieved the world's lowest energy consumption in that sector. In a "closed cycle" system of pulp production, hardly any water is discharged and elemental chlorine has been replaced with chlorine dioxide, hydrogen peroxide, oxygen, or enzymes.

In the United States, the paper industry meets roughly half its energy needs using not only wood but also sawdust, scrap wood, and pulping waste to fuel its boilers for heat and electricity generation. Following the introduction of the 1978 Public Utility Regulatory Policies Act — which requires electric utilities to buy power produced by independent companies at fair prices — the biomass-fueled grid-connected electricity capacity in the United States rose from 200 megawatts in 1979 to some 6,500 megawatts by 1994. In order to meet strict sulphur emission standards, American utilities have also shown interest in burning low-sulphur wood along with coal, with the result that the paper and wood-products industries in the United States have increased the biomass energy potential to 10,000 megawatts. However, some of the biomass converted materials

4

46. *In the remote Bunair Valley region of Swat, Pakistan, a newly received PV panel lies on a heap of biomass as young girls carry the straw for fuel in their homes fitted with solar lights.*

47. *Children in India are benefiting from the market-oriented biomass gasification program for water pumping in rural areas such as this, where more than 1,440 systems with an aggregate capacity of 21 megawatts have already been installed.*

such as paper have a higher energy value if they are recycled or reused rather than burned. Hence, burning biomass refuse is not always the best solution, as recycling paper saves two to four times as much energy as can be produced by burning it. In fact, waste incinerators in the United States are already experiencing fuel shortages due to successful recycling and waste-minimization efforts.

In the developing countries the use of biogas has improved the quality of life in rural areas. Primarily produced from cow dung, it is now used for lighting and the running of dual-fuel engines for providing mechanical energy for drawing water and other applications. The residual slurry from biogas plants is helping to enrich manure and increase agricultural yield. The setting up of biogas plants has improved the environment and the sanitation in households. Eye and lung diseases caused by kitchen smoke have been reduced, and women and children are spared the drudgery of collecting and head-loading heavy bundles of wood fuel every day. This helps in saving wood, and hence preserves the forests.

For the past 80 years, increasingly large

48

49

48. Biogas is now increasingly used worldwide as a cooking fuel that is clean, efficient, and convenient. The family-type plant used by this woman in Gujrat is one of more than 2 million units installed in India; and additional 200,000 such plants are installed every year.

49. Sun-dried cow dung is widely used in most Asian, African, and Latin American countries as cooking fuel, material for house building, etc. In India, homage is paid

to biomass energy in a ritual by making a drawing of the sun with cow dung, and prayers are then offered to it with an oil lamp made of kneaded wheat flour.

50. Against the background of Sûrya sun symbols painted on the façades of their homes, Nepalese women wash up after cooking with biogas in Bagmati Zone village. The accelerated denuding of forests has alarmed the authorities, who are now offering financial incentives to people using biogas plants.

quantities of livestock and human dung have been gasified in anaerobic digesters, especially in India and China. In India, two schemes for the promotion of biogas plants are being implemented at present — the National Project on Biogas Development (NPBD) and the Community/Institutional Biogas Plant (CBP/IBP) Program. Family-size biogas plants are promoted under the NPBD scheme, and centralized aid grants cover about 20 to 25 percent of the cost of the plants. Up to 1996, some 2.18 million family-type biogas plants had been set up, and a further 200,000 plants are added every year. Research and development projects are increasing biogas yield as new digestive material, high-rate biomethanation processes and other measures are being introduced. Under the CBP/IBP – Night-Soil-Based Biogas scheme, large biogas plants based on recycled organic waste, recycled human waste, and community biogas plants for economically weaker sections of society are being promoted. Plant sizes vary from 500 cubic feet to 3,000 cubic feet (15 cubic meters to 85 cubic meters), and the government provides about 50 percent of the cost for setting up such plants. So far, 640 CBP/IBPs have been set up, as night-soil-based biogas plants are now becoming acceptable in rural areas.

It is estimated that the Chinese have installed more than 20 million household digester systems since the 1970s. They have also found that in isolated rural areas, biogas from farm waste can be used to run generators that produce electricity at a lower cost than conventional oil-powered generators, and about 1,000 large farm-waste digesters have been built during recent years. At one such big chicken farm at Liu Minying village, Da Xing County, near Beijing, a large digester working on the manure of 100,000 chickens saves about 8,000 tons of coal energy annually and provides cooking biogas to some 900 people in 240 households. Cooperative societies are building more such large digesters all along the new highway from Beijing to Hong Kong.

Raw materials such as sawdust, straw, wheat chaff, and the like are often readily available, but they burn so quickly that they are not very practical as fuel sources. Briquetting involves compacting the raw material into "bricks" so that the energy content is raised to a level approximately equal to that of wood fuel. Raw materials such as straw and charcoal dust do not stick together when simply compressed, so a binder has to be added. Commonly used binding agents include starches from maize, wheat, or cassava, sugarcane molasses, tars, resins, glues, fiber, fish waste, and certain plants, such as algae. The most widely used

51

52

51. Straw is abundantly available in regions where grain is intensively cultivated. Technologies for the use of surplus straw and its energy utilization are being developed in several European countries.

52. Research is being conducted in many countries to obtain biomass energy from high-energy hybrid grasses with a short rotation period, and from other unconventional crops, such as napier grass, switchgrass, kenaf, flat peas, etc.

noncombustible binders are ash, clay, or mud. Some countries, such as South Africa, have experimented with briquetting coal dust, straw, and wheat chaff, but the problem is finding a cheap, available binding substance and an appropriate technology to do it economically.

Ways to reduce consumption and make better use of available wood continue to be investigated, while replenishing wood supplies is of major importance. In an attempt to reduce the consumption of wood, researchers have developed fuel-efficient wood stoves. Especially made for use in the developing countries, these well-insulated stoves are shaped so that the wood burns slowly. The heat is delivered to the pots and not wasted, and the stoves produce less smoke and ash than traditional cookers. In many poorer regions, a type of charcoal stove made of clay and a used oil drum has become a popular kitchen accessory as it consumes half the energy of traditional metal stoves. In East Africa this kind of ceramic stove is called the Kenya Ceramic Jiko, and more than a million of them have been produced over the past decade. The increasing demand and production has led to healthy competition among the local artisans who produce them, with the result that their prices have fallen and most families can afford to buy them.

Just as the energy economy of Inner Mongolia is dependent upon its vast treeless grasslands, Denmark — a country with extensive cropland but few large forests — has made straw into an important energy source. It is estimated that the surpluses of this agricultural waste product can meet more than 7 percent of the country's energy needs. With its decentralized approach, the Danish government has encouraged the construction of small-scale straw burners that provide heat for on-farm use. And since 1980, more than half of the district heating systems have been modified to rely on straw for an average of 90 percent of their fuel, while ash, a by-product, is being used by farmers to fertilize their fields. The government has also imposed higher taxes on fossil fuels, a move that is expected to boost straw combustion to more than 1 million tons by the year 2000. The use of livestock manure is also being encouraged, since it is calculated that oil imports to Denmark could be reduced considerably if all of the livestock manure in the country were used to produce biogas.

Bagasse, the crushed fibers that remain after sugar has been removed from sugarcane, is one of the largest sources of biomass energy in the world. Brazil, which has to import about half of its oil, produces ethanol from sugarcane.

53

54

53. The traditional use of passive heat for fish drying is being improved in many countries by employing other forms of solar energy with the potential to become part of the aquatic energy farms of the future.

54. Among the unconventional energy sources is fast-growing algae for the production of spirulina and the use of other marine plants, such as giant seaweeds, for ocean farming.

Pure ethanol produced from sugarcane molasses is now being used in a third of the automobiles manufactured in Brazil. This program, launched in 1975 by Brazil's Proálcool, has increased its annual ethanol production about twenty-fold from around 200 million gallons (720 million liters) when the program began to about 50 billion gallons (19.3 billion liters) per year by the mid-1990s. Other substances — such as wheat and other cereals, sugarcane, beets, wood chips and forest residues, kelp, macadamia nut husks, and many other land and sea plants — can also produce ethanol by the fermentation of biomass materials. Bagasse can be burned in conventional boilers to produce steam for the sugar extraction process, and also to produce electricity. A typical sugar mill boiler produces enough steam to fuel the plant's operation and generate electricity for external users. Improvements in the tech-

nology are being made in a project funded by the Global Environmental Facility in Brazil's northeast state of Bahia; in this plant 25 to 30 megawatts of electricity will be generated by turbines driven by gas produced from wood chips. Studies are being conducted in São Paulo to use bagasse.

Ethanol from sugarcane molasses is also manufactured in Hawaii and mixed with gasoline to produce fuel for automobiles. But the Hawaiian sugar industry is concentrating more on the production of electricity by burning bagasse, which is now sufficient for more than 60 percent of the electric power demand on the island of Kauai and about 45 percent of that on the island of Hawaii. More advanced steam turbines operating at higher pressures are already in use and can maintain steam production while raising electricity output three-fold. At present the electricity-generating capacity from bagasse

55

55. Biomass energy and biogas by-products are effectively used by the 300,000 herdsmen who live in the 1,100 villages of Inner Mongolia's vast grasslands. By 1996, more than 9 million household digesters of domestic and farm waste had been built in China as a whole.

is about 200 megawatts in the state of Hawaii, and the fuel provides another 80 megawatts in Florida and Louisiana. Similarly, the availability of inexpensive bagasse has encouraged sugar mills in India, Costa Rica, Cuba, Fiji, Guatemala, Mauritius, Thailand, Zimbabwe, and several other developing countries to produce and sell surplus electricity to local power companies.

Gas turbines can also be fired with other biomass residues, such as the forestry waste currently burned in paper and pulp factories. India — one of the world's leading producers of rice — is using rice husks to generate electricity in the state of Punjab, where one biomass facility is expected to generate more than 10 megawatts of electricity by burning rice straw. Each year, the country accumulates more than 20 million tons of rice husks, and if all of this waste material were used for power generation, this source alone could produce enormous quantities of electricity.

As with all other forms of sustainable energy, bio-fuels are viable only if they are regularly replaced. Trees need to be replanted and organic fertilizers need to be returned to the soil for the cycle to be continuous in the long run. Tree planting campaigns have been initiated in some countries, although frequently the problem is recognized and dealt with only after substantial shortages have already occurred. A study commissioned by the United Nations for the 1992 Conference on Environment and Development found that one way to increase biomass production is to cultivate fast-growing crops, especially for a particular use, and to replace them expeditiously. Some trees can be harvested every 2 to 8 years, compared to the normal span of 30 to 60 years between harvests in conventional forestry. Perennial grasses such

56

56. *The abundance of straw in Inner Mongolia is widely used by thousands of families for cooking and heating. It is also mixed with mud and used for heat insulation in homes.*

as switchgrass and elephant grass could also be grown and harvested every 6 to 12 months using improved methods similar to those used for growing alfalfa and other forage crops.

A good example is set by Finland, where the growth and endurance of more than 250 willow varieties have been tested in an area of 7 acres (2.8 hectares) at the Kopparnäs energy park. Willows are normally harvested every three or four years and the average annual willow growth is about 5 feet (1.5 meters). But with new methods using suitable substrate, the best species can grow as tall as 16 feet (5 meters). The research, which started in 1983, is centered on producing short-cycle crops that can grow rapidly and be harvested eight to ten times before the field is replanted. Harvesting methods, crop drying, chipping, and burning are also being investigated. Finland also has a good record in saving wood through paper recycling—half a million tons in 1996.

Several seaweed varieties, such as giant kelp — besides being a source of food, fertilizer, and chemicals — have the potential to provide substantial amounts of energy since they grow very quickly and can be harvested repeatedly. Experiments in the United States show that under favorable conditions kelp farms can yield an average of 20 tons of biomass per acre annually and this weight of kelp can produce more than 175,000 cubic feet (5,000 cubic meters) of methane per acre annually — about the same amount of energy as 1,850 gallons (7,000 liters) of oil. Theoretically, if all the energy stored each year in biomass could be made available for human use, it would provide about ten times the total amount of energy consumed by people worldwide.

Another approach adopted in the mid-1980s at research stations in Kenya and Nigeria found that by mixing corn and leucaena trees, crop production could be in-

57

creased substantially while producing at least 2 tons of wood per acre (5 tons per hectare) annually. Similar methods are being employed in China, which has more than 1.25 million acres (half a million hectares) of mixed crops and trees in the North China Plain. Other varieties of trees that grow best in mild to tropical climates may have even higher yields — eucalyptus trees, for example, can reach a height of about 16 feet (5 meters) after only six to eight years, and provide from 400 to 1400 cubic feet (11 to 40 cubic meters) of wood per acre annually.

But do these efforts have any chance of succeeding? For some 3 million years humankind lived in reasonable balance with the natural ecosystem, and although people lived on the plants and animals, the extent of their depredation was limited and reversible. But now there is simply not enough land or grain to sustain the world's exponentially growing population, which has already so badly damaged the ecological balance that biomass energy is barely renewable or sustainable. In many parts of the world firewood is in increasingly short supply as growing populations have either burned the trees as fuel or converted everything from the tiniest patch of green to large rain forests into treeless agricultural land without even space for grazing. The resulting fuel shortages have forced women and children in poor countries to spend much of their time collecting wood, while in many areas crop residues and animal dung — which could otherwise be valuable fertilizers — are burned in cooking stoves.

Traditional cooking methods often capture only about 8 percent of the wood's energy — a colossal wastage of biomass energy considering that more than 2 billion poverty-stricken people worldwide are obliged to use such primitive methods for their very survival.

To sum up, primary biomass productivity

---

57. In Europe, research is being done to improve the production of miscanthus, the best known energy crop, which is planted in springtime with a density of 4,000 plants per acre (10,000 plants per hectare). In favorable weather conditions the rate of sprouting can reach more than 90 percent, although the plants take several years to mature.

58. Animals such as sheep are one of the three food-related

parts of biomass nutrition — producers (photosynthesizing plants), consumers (ingesting animals), and decomposers (absorbing bacteria and fungi).

59. PV-powered drinking troughs for animals have become very popular in Europe. The energy generated powers a pump that conveys water from an irrigation ditch into the metal trough. A battery and control system ensures that the trough remains full, even during overcast weather.

58

59

(usually measured in units of energy, such as gram calories per square meter per year) represents the total energy available to carry on all biological activities if no organic matter is brought into the community from the outside. Organisms of the community are grouped by positions along food chains linked together into a community-wide food web in which many animal species feed upon a number of other species. Thus the major modes of nutrition are the three trophic (food-related) parts of the community — producers (photosynthesizing plants), consumers (ingesting animals), and reducers or decomposers (absorbing bacteria and fungi). During the growth period biomass draws carbon dioxide from the atmosphere, and in the burning process it returns the same amount of the gas to nature. Life on Earth is sustained by the ecological balance created by this closed circuit in relation to its environment.

60. Strawberry production using cold seawater for soil temperature control and plant watering through condensation is being studied at the Natural Energy Laboratory, Keahole Point, Hawaii.

61. Crane for transporting straw bales in the Danish straw heating plant in Hadsten. The heating plant has an installed capacity of 11 megawatts.

62. The combustion chamber is the principal part of a biomass heating plant. Underfeed stoker, grate firing, injection firing, or fluidized bed combustion are used, depending on the capacity of the plant and type of fuel.

Invisible Energy of the Wind

Mâui-akamai invented the first kite, but he could not fly it without the energy of the wind, *mana*, which an old priest, Kaleiiokû — who lived in the Waipi'o Valley, known for its winds and rain — jealously kept in a covered calabash. It was not easy to persuade the miserly Kaleiiokû to lend Mâui some wind power. But finally, when the old priest agreed and opened the wind-gourd *ipu-makan*, out rushed a gust of wind that carried Mâui's kite far up into the sky like a sailboat gliding over the sea's waves.

Just as poetry anticipated prose in the development of human language, the power of the wind's seasonal currents had already carried the sailboats of coastal inhabitants much farther across the sea long before the caravans of the inland traveler's ancillary draft animals, such as oxen, mules and camels, started on their journeys. At about the time Peking man discovered how to use fire, the "great fleet" of Mâui-akamai's Polynesian ancestors was being powered by the wind's kinetic energy and paddled with muscle power. They followed the traditional routes of the "dreaming songlines" boats of the Australian aborigines, who had crossed the deep strait that separates their vast island continent from the mainland of Southeast Asia some 40,000 years ago. They seemed to know through observation and experience about the force — which we now call the Coriolis — that deflects the winds to the right in the Northern Hemisphere and to the left in the Southern Hemisphere.

The sails of Egyptian boats began to capture wind energy on the Nile River (c. 3300 B.C.) — images of which appear in predynastic Egyptian rock carvings in Wadi Hammamat in the Nile Valley. The Phoenicians, master seafarers of antiquity, left more explicit records in bas-relief, cylinder seals, on coins, and even through a crude terra-cotta model of a ship, made sometime between 1580 and 1200 B.C. at Byblos, on the cost of Lebanon.

The single-sailed boats of the Etruscans in pre-Roman Italy date from about 475 to 450 B.C. By the 7th century, in northern Europe and along the Bering Strait, a new type of Viking ship had appeared in which a small piece of cloth had grown into a square sail. The European galleons of colo-

65

63. Miguel de Cervantes's widely read classic Don Quixote shows how large and important windmills had become in the early 17th century, and it gave rise to new expressions in European languages, such as "tilting at windmills." Lithograph by Gustave Doré (1832–1883).

64. Farohar, the winged-disk sun symbol at the Parsi Anjuman Fire Temple, Bombay.

65. Windmill (15th century) on the island of Mykonos, Greece.

nial times were developed from the Greco-Roman merchantman, which carried two small triangular topsails in addition to the mainsail. Even though it was essentially a sailing vessel, it could be rowed with muscle power when the wind failed.

It was the energy of the wind that brought about ancient trans-Pacific contact and maritime adventures of discovery through trade and cultural interaction among people in faraway regions of the world. As within Asia, trans-Pacific contacts enhanced the peoples' ability to make better tools for the production and storage of energy. The similarities between prehistoric Japanese pottery and Ecuador's earthenware of the Valdivia Phase are evidence that such contacts took place around 3000 B.C. Recent archaeological finds at a site near Monte Verde, in Chile — including remnants of dwellings with wooden frames and animal-hide roofs, tools made of stick and bone, and more than 700 stone tools — indicate that these tools of heliotechnology had been brought there by Asian migrants as early as 12,500 years ago. The Monte Verde archaeologists have found a second, deeper layer of putative human artifacts that may be as much as 33,000 years old, which suggests that Asians may have arrived in the Americas directly by the open sea and not only on land through the Bering Strait as has been speculated until now.

The sail had harnessed wind power from the dawn of civilization but it was only about 2,000 years ago that it was first used in the form of the windmill in China, Afghanistan, and Persia, mainly to draw water for irrigation and for grinding grain. Records show that in Persia, wind power from a vertical-axis machine was in common use in the 7th century — a technique later brought to Europe from Syria by the German crusaders in 1190.

Holland owes its very existence to the windmill, since until about 1000 A.D. the country was barely habitable — it consisted of marshes with small sluggish streams separated from the sea by a belt of dunes and wide tracts of land. The country was repeatedly ravaged by floods, and large parts of land disappeared underwater to form inland seas, such as the Zuider Zee, in about 1300. In the notorious St. Elizabeth's Flood in November 1421, thousands of men, women, and children drowned, as well as whole herds of cattle. Large numbers of inhabitants were driven out of their homes and had to take refuge on mounds at higher levels, where they used biomass energy by burning

66

67

66. *A replica of a thousand-year-old Viking ship, Roskilde Fjord, Denmark.*

67. *A painting of the Norwegian expedition to the North Pole, headed by Fridtjof Nansen, who used a wind turbine to produce electricity for lighting when the ship Fram was locked in the polar ice for three winters (1893–96).*

dried peat and wood from the neighboring forests. This repeated flooding led to the construction of dams to hold off the open sea, and windmills came to be utilized for the first time to drain inland pools and lakes. Between 1608 and 1612, the 10-foot-deep (3-meter-deep) Lake Beemster was drained with the help of 26 windmills working in two stages. But unfortunately, the Zuider Zee dike burst and the work had to be done all over again.

The oldest known document containing a reference to windmills in Holland is said to be the privilege granted in 1274 by Count Floris V to the burghers of the town of Haarlem. There are also records of windmills in other towns, such as in Amsterdam dating from 1336 and 1342, and in Utrecht from 1397. At first these were small mills, which were replaced from about 1526 by a *wipmolen* (hollow post mill) and then with the classic drainage mills. Their commercial use began with the construction of large corn mills and even larger industrial mills, such as the octagonal wooden smock windmill with a revolving cap. Records show that in the second half of the 16th century, smock mills with tail poles were in use as an oil mill in 1582, a paper mill in 1586, a sawmill in 1592 — and after 1600, windmills were constructed every-

where for a wide variety of jobs. In fact, two-masted wind carriages were running in the Netherlands in 1600, and a speed of 18 miles (30 kilometers) per hour with a load of 28 passengers was claimed for one of them. In 1760 the Swiss clergyman J. H. Genevois suggested mounting a small windmill on a cart-like vehicle — its power to be used to wind springs that would move the road wheels.

Meanwhile, British windmill construction was improved considerably by refinements to the sails and by the self-correcting device of the fantail, which kept the sails pointing into the wind. Spring sails replaced the traditional canvas rig of the windmill with the equivalent of a modern venetian blind, the shutters of which could be opened or closed to let the wind pass through or to provide a surface for its pressure. Sail design was further improved with the "patent" sail in 1807. In mills equipped with these sails, the shutters were controlled on all the sails simultaneously by a lever inside the mill connected by rod linkages through the wind shaft, with the bar operating the movement of the shutters on each sweep. With these and other modifications, British windmills adapted to the increasing demands on power technology.

68. The European galleon of colonial times was developed from the Greco-Roman merchantman ship, which carried two smaller triangular topsails in addition to the mainsail. Essentially a sailing vessel, it could be rowed with muscle power when the wind failed. This picture of a Dutch galleon was painted by Ludolf Backhuyzen (1631–1708). Courtesy: Louvre Museum, Paris.

Thereafter, for more than two centuries, it was the power of wind energy that brought the sailing vessels carrying heavy timber from the countries on the Baltic Sea and that sawed the wood in the *Paltrokilometresolens* — the windmills that derive their name from the German Mennonites who wore flaring "palts-rokken" coats. It was wind energy too that powered the many industrial oil mills and paper mills. More than 50,000 windmills were at work in Europe in 1850, before they began to be replaced with the atmospheric steam engine, invented by Newcomen in 18th-century England. From then on, their number progressively declined and by the turn of the century no more than 10,000 windmills were still running in Europe. The traditional windmill suffered another setback when industry began to use James Watt's new condensing low-pressure steam engine, which was cheaper and more efficient than the Newcomen engine. As a result, the use of wind power declined sharply in the 19th century as steam spread and the scale of power utilization increased. Windmills that had satisfactorily provided power for small-scale industrial processes were unable to compete with the production of large-scale steam-powered mills. With this began the Industrial Revolution.

Steam did not simply replace other sources of power: it transformed them, with the result that windmills also improved in design and efficiency. At the same time the nostalgia for the good old windmill revitalized interest in renewable energy, and this led to the building of some 1,300 new wind turbines in 1916, to power threshing machines, grinding mills, and water pumps. The father of modern wind technology for electrification in Denmark was the

69. Construction of windmills improved considerably when spring sails replaced the traditional canvas rig with the equivalent of a modern venetian blind, the shutters of which could be opened or closed to let the wind pass through or to provide a surface for its pressure. Another innovation was the self-correcting device of the fantail, as seen in the Mediterranean windmills still commonly used in Majorca, Spain.

renowned physicist and meteorologist Paul la Cour, a teacher of natural science at Askov Folk High School from 1878 until his death in 1908. Starting almost from scratch, la Cour, in cooperation with two Danish engineers, Vogt and Irminger, laid the foundation of the industry. Over the years, the capacity of individual wind turbines has increased from those rated 15 to 22 kilowatts, to the larger 500 kilowatts, with 750- to 1,000-kilowatt turbines now in the pipeline. Today nearly all of Denmark's 3,700 wind turbines, which had a total installed capacity of 500 megawatts in 1994, are connected to the grid. Danish manufacturers of wind turbines cater to about one-third of the world market, and the country accounted for about 80 percent of Europe's installed wind capacity in 1996. The government has fully committed itself to the production of wind power and hopes that by

2005, 10 percent of Denmark's electricity consumption will be met from wind power alone.

Among the leaders in wind energy production is the United States. In 1925 the Jacobs brothers started producing battery charging wind turbines in the 2.5- to 3-kilowatt range, followed by the Aermotor Company in Chicago, which produced some 80,000 windmills, mostly for water pumping. The industry was greatly boosted as a result of the 1978 publication of a state government study that identified three large windswept passes in California's Coast Ranges — at Altamont Pass near San Francisco, in the Tehachapi Mountains north of Los Angeles, and in the San Gorgonio Pass near Palm Springs. Financial incentives offered by the government started the "wind rush" — similar to the one that followed the discovery of gold in the Sierra Nevada in 1848 — so that

70. "Mill language" is the message that a particular position of a mill's sails conveys — (a) celebration, (b) mourning, (c) short rest, (d) long rest. Illustrations by G. Pouw, Zaandam, Netherlands.

71. Poldermolen "De Kager" decked in colorful flags. Dutch windmills are cultural centers for the celebration of festivals, weddings, and other social gathering — a tradition that has spread throughout Europe.

72. (following pages) Windmill in the town of Wijk Bij Duursted on the river Lek. Oil on canvas by Jacob Ruysdael (1628–1682). Courtesy: Rijksmuseum, Amsterdam.

about 12,000 wind turbines were installed in California between 1982 and 1992 with a total generating capacity of more than 3,000 megawatts by the end of 1996. This represents enough energy to meet the residential needs of a city of about 1 million people. The most efficient turbines are able to convert about 35 percent of the wind's kinetic energy into electricity.

Large grid-connected wind turbines are most productive when — as with the ancient windmills — they are operated on a cooperative basis among the users and the utility services. The extraordinary success of Danish wind energy is largely due to its cooperative structure. During the late 1970s, the government established a wind turbine test station and introduced a 30 percent subsidy to reduce investment costs — although this has been phased out over a period of ten years. Since then, 78 percent

of wind turbines have been installed by independent owners, mostly rural cooperatives formed by groups of 30 to 40 families. More than 100,000 Danish families are now members of wind cooperatives (or wind turbine guilds), and many are also united in a national consumers' association of wind turbine owners. The cooperatives hold their general assembly once a year and also organize other social or cultural events.

The success of the Danish cooperatives and similar incentives provided in the development of California's wind farms are now being followed up in many parts of the world. There are three basic types of cooperative ownership. In the first, a group of people in a cooperative society buy a turbine with a loan from a bank or credit association, and each person or association of individuals finances a share equivalent to their electricity consumption. The electricity is then

73. Windmills greatly influenced European painting and its developing styles, as seen in the masterpieces of the Renaissance artists, the 16th- and 17th-century Flemish windmill paintings, such as those of Bruegel, the expressionists such as Van Gogh, and abstract painters such as Mondrian. This famous drawing on paper of a Dutch windmill is by Rembrandt (1606–1669).
Courtesy: Rijksmuseum, Amsterdam.

sold to the utility at a certain fixed price determined by the government. The second method is the outright purchase of a wind turbine by an individual farmer on whose field the turbine is installed. In the third category, wind turbines are paid for by a utility and installed in clusters — so-called wind farms. The turbines are usually grid-connected and the power can be sold to a power utility company or the government.

Taking its cue from the above, the central government of India has offered a number of fiscal incentives for development of this sector, which, supplemented by additional subsidies provided by the state governments, has resulted in the establishment of several joint ventures with foreign companies. In the windswept valleys of the Muppandal and Perungudi regions of Tamil Nadu, for example, joint ventures with Danish, Dutch, German, British, and American

companies have installed more than a thousand large and small wind turbines that produce a total of about 850 megawatts of grid-connected electricity. As compared with China's 56 megawatts of wind electric power, India has now over 900 megawatts with a potential of 20,000 megawatts in the country as a whole — making it the third largest wind energy program in the world.

Hybrid systems may connect wind turbines to one or two power sources. The wind/diesel combination is the most commonly used hybrid system, with the advantage that investment in the wind turbine can be paid for by the savings in fuel for the diesel generator. At Esperance on Western Australia's southern coast, diesel hybrid systems at two wind farms supply up to 17 percent of the town's electricity requirement. The wind farms are connected to the Esperance

74. An old Dutch-style windmill in Denmark, a country that is now leading in the manufacture of new wind turbines of the kind seen here in the background.

diesel power station, from where power is distributed on the regional grid. The first wind farm was set up at Salmon Beach in 1987 with six wind turbines, generating a total of 360 kilowatts, and they contribute about 2 percent of Esperance's power needs, saving about 66,000 gallons (250,000 liters) of fuel annually. The success of this project led to the establishment of a bigger wind farm close by at Ten Mile Lagoon. Here, nine 225-kilowatt wind turbines were installed at a cost of about 5.8 million dollars for a total capacity of 2,000 megawatts. The scheme supplies up to 15 percent of the town's electricity needs and saves about half a million gallons (2 million liters) of fuel a year. It is expected to recover its building costs in about 12 years. A computer at the power station controls the operation of the turbines — some 9 miles (14 kilometers) out of town — and the operation of the

diesel generators that respond to wind variability and maintain voltage levels.

While large computerized grid-connected wind farms are economically justifiable for supplying energy to urban areas, the only practical way of providing electricity to the isolated countryside is through the installation of much smaller, stand-alone eolic-photovoltaic systems. They have an advantage over the diesel-only or photovoltaic-only systems that are widely used for power generation in the world's remote regions. Such renewable energy systems are being increasingly installed not only in the developing countries but in industrialized regions of the world as well. A Spanish cattle farm in the Alt Urgell district of Catalonia, for example, has a self-generating installation using a 5.6-kilowatt photovoltaic system with a 1-kilowatt aerogenerator. The Argestues dairy

75

76

75. *The popularity of windmills in Holland during the 16th and 17th centuries is seen in this engraving by J.B. Collaert (1566–1628).* Courtesy: Rijksmuseum, Amsterdam.

76. *The drainage windmill Trouwe Wachter (1832) at Tienhoven, Utrecht, is decorated with flags and bunting to celebrate the 25th anniversary of the miller's wedding.*

farm with about 50 milk cows has been using a 15-kilowatt diesel generator for the past 15 years. But since March 1992, it has been equipped with a hybrid eolic-photovoltaic system that has considerably reduced the diesel fuel costs. Similar small hybrid solar energy systems have since been installed at six other isolated homes in the municipality of Vallis d'Aguilar. Small eolic systems are now being used at tens of thousands of remote sites worldwide to power communications equipment, scientific instruments, and monitoring equipment. At the South Pole, where harsh weather conditions and six months of darkness severely limit the available power options, energy from the wind has been used for more than 50 years.

77

China has pointed the way to how the common people's energy problem could be solved through passive solar heated houses fitted with small wind-only or wind-photovoltaic hybrid systems. In the remote areas of Inner Mongolia's vast grasslands — where around 300,000 herdsmen live in 1,100 small villages — some 120,000 homesteads have been equipped with 125,000 sets of government subsidized stand-alone wind generators and photovoltaic systems over the past five years. By the end of 1996 an estimated 3,500,000 cubic feet (100,000 cubic meters) of housing and 150 million cubic feet (4.3 million cubic meters) of livestock sheds had been provided with solar electricity or equipped with passive heating. There are now about 100 manufacturing plants in China producing such units, of which 17 are making a range of products in Inner Mongolia alone. They produce

78

small wind generators (5 to 50 kilowatts) and wind pumps and provide training facilities for local inhabitants.

Shangdu Husbandry Machinery Factory is one of the 10 biggest producers of small eolics in the world, exporting its products to several foreign countries. Renewable energy is on the curricula of many schools, and some 20 research institutions employ more than a thousand scientists working in the field.

Interest is now increasingly centered on "sea-based wind farms" — groups of wind turbines connected by undersea cables to the power grid onshore — which could generate electricity at many coastal locations. In Denmark the first offshore wind farm was inaugurated at Vindeby in 1992 and several more are in the planning phase. According to the official Danish "Energy 2000" plan, offshore wind power will supply 10 percent of the country's power demand by 2005 — equivalent to an installed capacity of 1,500 megawatts. Many countries with extensive coastlines could obtain most of their power from wind farms located on offshore platforms in shallow seas.

It is estimated that the world had roughly 25,000 wind turbines in operation at the end of 1996, with installed capacity of about 6,000 megawatts of electricity — 40 percent of it in California and Denmark. The European Union has been helping countries such as Spain, where several wind energy projects have been set up, such as the 20-megawatt Pesar, 5-megawatt Los Valles, 0.65-megawatt Levantera, and 20-megawatt Juan Grande wind farms. It has plans to

---

77. *A statue of Vayu, the Vedic wind deity, was recently unearthed near the 12th-century Raja-Rani temple in Bhuvaneshwar, India.*

78. *A farmer in Thailand working in his rice field, irrigated with water drawn from a well by wind power.*

79-80. *Unconcerned with the more than a thousand large and small wind turbines that have been installed in the windswept valleys of the Muppandal and Perungudi areas of Tamil Nadu, India, a woman goes out to graze the cattle while her husband works at home.*

79

80

82

83

install 4,000 megawatts of wind power capacity by 2000, and 8,000 megawatts by 2005 — more than six times the level in 1993. Europe as a whole could obtain a quarter of its total electricity requirements from the wind, while other countries, such as Mexico, Egypt, Tunisia, India, and South Africa, should easily be able to extract one-fifth of their electric power from the wind. Countries with considerable wind potential include Argentina, Canada, Chile, China, Russia, Spain, and Great Britain, and the numerous small subtropical island countries that have nearly constant trade winds.

People all over the world are turning to wind energy not only because it is cost-effective, clean, and safe, but also because windmills have traditionally been a part of their cultures and traditions. They are emotionally integrated into the customs and ceremonies of people's joys and sorrows, especially in Europe. This influence can be seen in the developing styles of European paintings — the masterpieces of

the Renaissance artists, the 16th- and 17th-century Flemish paintings of windmills by Bruegel and the sketches of Rembrandt, the expressionism of Van Gogh, and the abstract paintings of Mondrian. The adventures of Don Quixote, who, mistaking windmills for giants, attempts to joust them, inspired one of the most prolific and successful French book illustrators, Gustave Doré (1832–88).

Miguel de Cervantes's widely read classic *Don Quixote* shows how large and important windmills had become in the early 17th century and gave rise to new expressions in European languages, such as "tilting at windmills." During World War II, the position of a windmill's sail was frequently employed as a signal to transmit secret messages and prayers for the safety of the partisans in hiding; and even today windmills are decked with flags and bunting, and the different positions of the sails announce festivals, weddings, funerals, and other social events.

81. An illuminated wind turbine in California's windy Altamont Pass conveys season's greetings on behalf of some 12,000 wind turbines that had a total generating capacity of more than 3,000 megawatts by the end of 1996.

82-83. Wind energy projects have been set up across several wind- swept passes in Spain, including Tarifa. Plans are to install 4,000 megawatts of wind power

capacity by the year 2000, and 8,000 megawatts by 2005 — more than six times the level in 1993.

84. (following pages) Wind turbines installed on the seashore of Århus Harbor, Denmark. Nearly all of the country's 3,700 wind turbines with a total capacity of 500 megawatts are grid-connected, and 78 percent of them are operated by independent, mostly rural cooperatives.

85

86

87

85. *Wind-diesel power — with which Neste is experimenting along the Kopparnäs seashore — is a competitive alternative for electricity generation in isolated areas in Finland.*

86. *In Denmark the first offshore wind farm was inaugurated at Vindeby in 1992, and several more are now in the planning phase.*

87. *The offshore wind farm near Vindeby is the precursor of the Danish "Energy 2000" plan, which will supply 10 percent of the country's power demand by 2005 — equivalent to an installed capacity of 1,500 megawatts.*

88

89

88. Wind turbines at Blyth Harbour, Northumberland, England — a European Community THERMIE project.

89. Scientists measure wind speed in Antarctica's Wright Valley to assist the modeling of climate processes in the McMurdo Sound dry valleys.

90. following pages) In one of the remote areas of Inner Mongolia's vast grasslands, a herdsman's family stands proudly in front of their yurt with some of their 500 sheep. Their yurt is one of the 120,000 homesteads that have been equipped with 125,000 sets of government-subsidized stand-alone wind generators and PV systems during the past five years.

Transparent Energy of Water

Mâui-akamai discovers the tremendous power of water when the jilted suitor of his mother, a giant eel named Kuna, used it to intimidate her first by stopping the water in the Wailuku River and then by flooding her cave — and Mâui defeated Kuna by using the even greater power of steaming water by throwing burning volcanic stones into the river. "Water gave birth to Agni" — energy of fire — states the *Rig Veda*, and the sun god Vishnu is called Sûrya Narayana, or "the Sun Deity Moving in the Waters," as he sleeps on the serpent of darkness. The Egyptian theology of Memphis has it that Ptah is both the male and female primeval ocean that gave birth to the sun god, Atum. According to a Chinese myth, when the strongman Kun failed to build a dam and stop the floodwaters from causing widespread destruction to life and property, his son Yü adapted a new technique of making canals for the water to drain off to the sea. As a reward, he was elevated to the throne and founded the Hsia Dynasty.

The idea that the waters of the Earth undergo cyclical motions, changing from sea water to vapor to precipitation and then flowing back to the ocean, is probably older than any of the surviving texts that hint at or describe this cycle explicitly. Xenophanes held that the source of precipitation is to be sought in the sea, which is "the begetter of clouds and winds and rivers." Aristotle was more specific when he wrote in his *Meteorologica* that by "the sun's heat the finest and sweetest water is every day carried up and is dissolved into vapor and rises to the upper region, where it is condensed again by the cold and so returns to the Earth. ... So the sea will never dry up; for before that can happen the water that has gone up beforehand will return to it."

Although the origin of the water in the Earth that seeps or springs from the ground was long the subject of much fanciful speculation, the arts of finding and managing water in the ground were already highly developed as far back as the 25th century B.C. Mesopotamian clay tablets make frequent reference to canals or reservoirs supplying cities between the Tigris and Euphrates rivers. The Letters of Hammurabi (c. 1760 B.C.) refer to the cleaning of canals. The Minoans, who flourished on Crete (c. 2000–1400 B.C.), had highly developed sanitary facilities, flushed with water from an aqueduct system. The great Assyrian aqueduct of Jerwan was built around 691 B.C. under Sennacherib, and joined earlier systems to bring water from a tributary of the Greater Zab to Nineveh, some 50 miles (80 kilometers) away.

The construction of long, hand-dug underground aqueducts known as *qanats* in Armenia and Persia represents one of the great hydrological achievements of the ancient world. *Qanats* derive their water from permeable deposits of

---

91. In the holy waters of the Ganges River, a father and son pay homage to the rising sun. "Water gave birth to Agni —the energy of fire," states the Rig Veda, and in most ancient civilizations the sun is represented on Earth by fire.

92. Varuna is the oldest Vedic deity of water. As the regent of lunar mansions, he is represented with a moon halo. He makes the sun shine, winds blow, and creates channels for the rivers to flow into the ocean.

gravel and sand washed down by streams from mountain ranges into drier lowlands along their borders. These tunnels are commonly 6 to 10 miles (10 to 16 kilometers) long and can be up to 400 feet (120 meters) below the surface. After some 3,000 years, *qanats* are still a major source of water in Central Asia where an estimated 20,000 *qanats* are still in operation today.

The *qanat* developed into the aqueduct, and between 312 B.C. and A.D. 226 the Romans completed 11 aqueducts, mostly underground but partly on arches, and fed from springs between 10 and 60 miles (15 and 90 kilometers) from the city of Rome. These aqueducts were gravity-fed, low-pressure watercourses, designed primarily to serve urban needs

rather than those of the farm. The water gradually descended some 800 feet (250 meters) through a series of free-flowing conduits to distribution tanks, of which there were 247 in the city in the 1st century A.D. At these tanks, the system changed from one of gravitational flow in conduits to one of a low-pressure supply in pipes. In the building of conduits, trenching was employed where possible, and tunnels — which could be 50 feet (15 meters) or more underground — had shafts at intervals of 250 feet (75 meters) or so to prevent air locks and to permit inspection and cleaning. The conduits and pipes, depending on their time and place of construction, were made of a variety of materials — stone-built ducts, open ducts of

---

93. Landscape with Farm and Watermill. A lithographic print by Jan Dam Steuerwald, after M. Hobbema. Courtesy: Rijksmuseum, Amsterdam.

94. Children line up against the animal yoke used for drawing water from a bucket-chain well in Muskipur village, Bunair Valley, Pakistan.

masonry, fittings of bronze, pipes of stone, terra-cotta, wood, leather, and lead. In the older aqueducts, such as Appia, the channel was lined with cut stone walling made of a friable gray tufa called *capellacio*. Tufa is porous rock formed as a deposit from springs or streams.

The power of gravity and pressure to siphon water through a pipeline or a tunnel may have germinated the idea of the water mill. The Byzantine general Belisarius is credited with having invented the first floating water mill in 536 A.D. The Greeks and Romans also began using water power, as is evident dating back to the 1st century B.C. from the writings of Greek historians who describe how waterwheels were used to grind corn. The type of water mill

that flourished first in northern Europe appears to have been the Norse mill, which used a horizontally mounted waterwheel driving a pair of grindstones directly without the intervention of gearing. Examples of this simple type of mill survive in Scandinavia and the Shetlands; it also cropped up in southern Europe, where it was known as the Greek mill. In England, many of the 5,624 mills recorded in the Doomsday Book — a 1086 census of English landowners and their property — were probably of this type, although it is likely that by this date the vertically mounted undershot wheel had established itself as more appropriate to the gentle landscape of England. Its use extended to fulling cloth (shrinking and felting woolen fab-

95. Water mill in the series "Nova Reperta".
Lithographic print by Th. Galle/ H. Collaert.
Courtesy: Rijksmuseum, Amsterdam.

96. A Pakistani farmer draws water from a well with his bucket- chain device of the kind still commonly used in Asian, African, and Latin American countries, where agricultural production has increased since the farmers started using greenhouses made of plastic material, as seen in the foreground.

97

98

rics), sawing wood, grinding grain, pumping water, crushing vegetable seeds for oil, and powering simple machines such as bellows.

The Romans are not known to have built any outstanding dams in Italy, although one would imagine that they needed large reservoirs of water to feed their numerous aqueducts. The two Roman dams still in use in southern Spain, at Proserpina and Cornalbo, show that more attention was paid to the Roman colonies. The 40-foot-high (12 meters-high) Proserpina Dam has a masonry-faced core wall of concrete backed by earth — it may therefore be regarded as a forerunner of the modern earth dam. The Cornalbo Dam represented a further advance in design with its masonry wall constructed of cells filled with stones or clay and faced with mortar. It differs from Proserpina in having a sloping upstream face and being straight in plan. The merit of curving a dam upstream was apparently not fully appreciated by the Romans until one such curved structure, the forerunner of the modern "arch-gravity" dam, was built in 550 A.D. at Dâra on the present Turkish-Syrian border by Byzantine engineers.

Until hydroelectric power was discovered, dams were structures built across a stream, river, or estuary in order to collect water for

human consumption, irrigation, to control floodwaters, or to increase the depth of river water for navigation in lean periods. Most dams continue to be of two basic types: masonry (concrete) and embankment (earth fill). Masonry dams are typically used to block streams running through the sort of narrow gorge found in mountainous terrain; though such dams may be very high, the total amount of material required is limited. Embankment dams are preferred to control broad streams, where only a very large barrier requiring a great volume of material will suffice. The dam contains pipelines (penstocks) or tunnels equipped with valves or gates that can control the rate of water flow. When these structures are opened, the stored water flows down from the higher level behind the reservoir to the lower water level downstream of the dam. The rushing water — which the Greeks used to "siphon" — is usually channeled through turbines coupled to generators near the bottom of the dam to produce electricity.

The earliest recorded dam is believed to have been on the Nile River at Kosheish, where a 50-foot-high (15-meter-high) masonry structure was built around 2900 B.C. to supply water to King Menes's capital at Memphis. Evidence exists of a masonry-faced earth dam built about 2700 B.C. at Sadd-el-Kafara, 18 miles (30 kilome-

97-98. "Opwettense Watermolen" (c. 1743), a water mill for grinding grain, at Opwetten-Nuenen, Holland.

99. Broekmolen, a drainage mill (1581) at Streefkerk, Holland.

101

102

ters) south of Cairo; this dam failed shortly after its completion when, in the absence of a spillway, it overflowed during a flood. The oldest dam still in use is a rock-fill structure about 20 feet (6 meters) high on the Orontes Nahr al-'Asi in Syria, built about 1300 B.C. The Assyrians, Babylonians, and Persians all built dams between 700 and 250 B.C. for water supply and irrigation. Contemporary with these was the earthen Sudd-al-Arim Dam — 45 feet (14 meters) high and nearly 2,000 feet (600 meters) long — built near Marib in the Yemen. Flanked by spillways, this dam delivered water to a system of irrigation canals for more than 1,000 years. Other dams were built in this period in Ceylon (modern Sri Lanka), India, and China.

In 240 B.C. a stone crib 100 feet (30 meters) high and about 1,000 feet (300 meters) long was built across the Gukow River in China. Many earth dams of moderate height, and in some cases of great length, were built by the Sinhalese in Ceylon after the 5th century B.C. to form reservoirs or tanks for extensive irrigation works. The Kalabalala Tank — formed by an

earth dam 80 feet (24 meters) high and nearly 4 miles (6 kilometers) in length — had a perimeter of 37 miles (60 kilometers) and helped store monsoon rainfall for irrigating the country around the ancient capital of Anuradhapura. Many of these tanks in Ceylon are still in use today. In Iran, the Kebar, a pioneering arch dam, was built early in the 14th century. Spanning a narrow limestone gorge, it reached 85 feet (26 meters) in height with a thickness of less than 16 feet (5 meters). The central curved portion, 123 feet (38 meters) in length, was supported on two straight abutments. In Japan the Diamonike Dam was built to a height of 100 feet (32 meters) in 1128 A.D. Numerous dams were also constructed in India and Pakistan. In India, a design employing hewn stone to face the steeply sloping sides of earth dams evolved, reaching its zenith in the 10-mile-long (16-kilometer-long) Veeranam Dam in Tamil Nadu, built between 1011 and 1037 A.D.

Following a long period of inactivity during the Dark Ages, dam construction resumed in the 15th century and continued unabated until mod-

100. Windmill in Sunlight by Piet Mondrian (1872–1944), the kind of mill used for water drainage.
Courtesy: ABC— Mondrian Estate Holtzman Trust, New York.

101. Wind power is now used worldwide for pumping water in remote areas such as the vast desert of central Australia.

102. A farmer and his wife drawing water from a well underneath one of the 19 wind turbines installed at the Bulak Baru hamlet in the Jepara region of Indonesia.

103

ern times. Among these later dams was the Tibi (1579–89) in Spain, an arch-gravity structure 140 feet (42 meters) high — a height not surpassed in western Europe until the building of the Gouffre d'Enfer Dam in France almost three centuries later. In Europe, where rainfall is ample and well distributed throughout the year, dam construction before the Industrial Revolution was on only a modest scale and was restricted to forming water reservoirs for towns, driving water mills, and making up water losses in navigation canals. Then, as knowledge of the properties of materials and structures increased, engineers such as W.J.M. Rankine, professor of civil engineering at Glasgow University, provided a better understanding of the principles of dam design and the performance of structures. This led in turn to improved construction techniques and larger dams.

103. The Itaipu Hydroelectric Power Plant is a joint venture between Brazil and Paraguay. At present it is the largest dam in the world at 4.8 miles (7,760 meters) in length and 620 feet (190 meters) in height. The even larger Three Gorges Dam is being built in China, which will block a 370 mile (600 kilometer) stretch of the fast-flowing Yangzi River from Chongqing to Yichang at a cost of more than $24.5 billion by its completion in 2009.

Water power entered a new era in the late 19th century with the development of the world's first hydroelectric power plant in Appleton, Wisconsin, in 1882. Its 12-kilowatt power station produced enough power to light about 250 electric lights. The number and installed capacity of hydropower plants increased rapidly in industrialized countries during the late 19th and early 20th centuries.

Between 1950 and 1995 worldwide hydropower production increased more than seven-fold, rising annually from about 340 billion kilowatt-hours of electricity per year to 2.5 trillion kilo-watt-hours — providing just under one-fifth of the world's electricity. Although major hydropower facilities often require large initial investments — in the hundreds of millions to billions of dollars — these costs are more than

104. *When completed, the Canadian La Grande Complex hydroelectric project will be the world's biggest dam project ever. Started in 1973, it is being built in northern Quebec, diverting three rivers, reversing the flow of another, and then channeling the water from those four rivers into the La Grande-Rivière, which flows into James Bay.*

offset over the lifetime of the plant because hydropower facilities require no fuel, have lower operating costs than fossil-fuel or nuclear facilities, and may provide power for a long period of time.

One of the oldest methods of storing large quantities of water still used in hydropower technology is known as the pumped storage facility, which has two reservoirs at significantly different elevations. When baseline power plants — such as coal-fired or nuclear facilities — generate more electricity than needed, the excess power is used to pump water from the lower reservoir to the upper reservoir of the storage plant. And when more power is needed to supplement the power produced by baseline plants, the stored water in the upper reservoir is released and channeled through turbines to the lower reservoir to generate electricity. Since 1929, when the world's first 32-megawatt pumped storage system was built by the Connecticut Light and Power Company, this method of storing energy has been widely used in more than 400 installations worldwide, primarily in western Europe, the United States, and Japan.

105. Huge volumes of water require equally large turbine shafts. At Itaipu, they measure 8.5 feet (2.6 meters) in diameter, 18 feet (5.5 meters) high, and weigh 130 tons.

106. The main structural components of the Itaipu plant are the earth-fill and rock-fill dams, the power house, the diversion channel, the water power intakes, and the spillway.

The Aswan Dam in Egypt was one of the first gigantic hydropower projects that highlighted both the positive as well as the negative consequences of building large hydropower projects. Even though the 360-foot-high (111-meter-high) dam with a crest length of 2.4 miles (3,830 meters) and a volume of some 1,500,000,000 cubic feet (42,600,000 cubic meters) of water has helped in controlling the devastating annual flooding of the Nile River, many wonder if its environmental and cultural repercussions — the uprooting of large populations and the displacement of the famous monuments in the Nubian Valley — justified the billions of dollars spent on its construction.

Another gigantic project, the Itaipu Binacional hydropower station on the Paraná River bordering Brazil and Paraguay, involved the construction of a dam with a total length of 25,260 feet (7,772 meters) and an elevation of 740 feet (225 meters) above sea level at its crest. Completed in 1991, this 18-billion-dollar power station has eighteen 700-megawatt generators, which already provide a total of 12.6 million

kilowatt-hours — greater than other comparable dams, such as Guri in Venezuela (10.3), Grand Coulee, U.S.A. (6.5), Sayano Shushenskaya, Russia (6.4), and Krasnoyarsk, also in Russia (6.0). The first of its units began producing electricity in October 1984, and by the end of 1995 Itaipu had produced 500 billion kilowatt-hours of electric power in 11 years of commercial operation, providing 79 billion kilowatt-hours of electricity annually to locations throughout Paraguay and to industrial areas in southern and southeastern Brazil, including the nation's two largest cities, São Paulo and Rio de Janeiro. The complex, which has the highest hydropower potential in the world at around 40,000 megawatts, is expected to produce an additional 500 billion kilowatt-hours in the next 7 years.

An even larger hydroelectric project is the Canadian La Grande Complex being built in northern Quebec, construction of which started in 1973. The project involved diverting three rivers, reversing the flow of another, and then channeling the water from those four rivers into the La Grande-Rivière, which flows into James

107-108. The environmental benefits of small hydros such as the one at Aberdulais Falls at Neath in Wales have led to their increasing use in industrialized countries — Sweden produces 8,400 megawatts of electricity from some 1,350 plants; the United States has 1,700 plants with 3,420 megawatts; Italy's 1,400 plants have 1,969 megawatts capacity; and 1,500 plants in France generate 1,646 megawatts.

109

110

111

112

109-111. In China, some 60,000 small hydros, generating 13,250 megawatts, provide a sustainable source of electricity for rural communities. Small turbines are specially manufactured and installed at a number of mini-waterfalls built along canals fed by large reservoirs.

112. The World Bank's evaluation of 31 developing countries found that at least 28 of them had programs to develop their small hydro resources.

Bay. Three large power stations — La Grande 2, 3, and 4 — with a combined generating capacity of about 10,282 megawatts had been completed by 1986, and four other plants built on the La Grande, Brisay, and Laforge rivers have since increased the capacity of the complex to 14,960 megawatts. The builders, Hydro-Quebec and SEBJ, have also filed permits to add two more hydropower plants with a combined capacity of 785 megawatts to the La Grande Complex. In addition, the utilities have planned two other major hydropower projects in northern Quebec: the Great Whale Complex and the Nottaway-Broadback-Rupert Complex. However, the future of these new projects is in doubt because of their adverse environmental impact on local inhabitants.

Opposition to gigantic dams is mounting worldwide, primarily because they often adversely affect the cultural and environmental equilibrium of the region in which they are built. For example, recent studies of the La Grande reservoirs have found high mercury levels in both water and fish. Scientists believe that the flooded soils and decomposing vegetation release a soluble form of mercury that accumulates in certain fish species. Hence, the project is opposed by native Indian tribes who live on fish that may accumulate high levels of the toxic metal. Moreover, the reservoirs have flooded thousands of square miles of natural wilderness that was home to numerous animals, including black bears, polar bears, moose, lynx, and many varieties of ducks and migrating birds. The tribes have also claimed that industrial development associated with the La Grande Complex has interfered with their traditional lifestyle and is the cause of increasing social problems, such as drug abuse. In response to a suit by the native Indians, a Canadian court has obliged the Canadian government to conduct additional environmental impact studies to determine whether or not the Great Whale project should proceed.

Public discontent has been voiced in other countries as well. A Malaysian court has ruled that the 5.5-billion-dollar Pergau hydroelectric dam to be built in a Borneo rain forest ignores tribal sentiments and violates environmental law. The opposition intensified when it was found that British aid for the project was tied to huge arms sale to Malaysia. In Japan, the Ainu people have protested against the test-pumping of water into the Nibudani Dam on the Saryu River at Hiratori, Hokkaido, because the dam will submerge a holy site of the Ainu

113. Millions of people worldwide live on freshwater fishing and permanently reside on houseboats— principally in the delta areas of eastern Asia, where dwelling, commerce, and travel can be done almost exclusively on water. Their lifestyle has changed for the better since these literally floating populations began using photovoltaic systems for electricity. This Vietnamese fisherman owns a number of electrical appliances powered by a 30-watt PV panel installed on the roof of his boat, in addition to a solar lantern he uses for night fishing.

community. In a ritual they offered prayers to the Ainu god of fire and apologized to the sun god for not having been able to save their holy temple from submersion. In India there is concern about the damage that the building of Sardar Sarovar Dam on the Narmada River would cause to traditional cultures and historical monuments. About a hundred historic sites are likely to be submerged by the reservoir — the main reasons why the World Bank withdrew its financial contribution for the construction of the dam.

The Chinese have adopted a dual approach. While announcing plans to develop 11 large hydropower projects by the year 2000 — each with a capacity of more than 1,000 megawatts — they are also concentrating on building so-called "small hydros." Among the largest of the former projects is the Three Gorges Dam on the Yangtze River at Jinsha Jiang in Hubei Province, which is expected to be about 1.43 miles (2,309 meters) long and 600 feet (185 meters) high, forming a huge reservoir. It will have a generating capacity of 18,200 megawatts when completed in about 15 years time at an estimated cost of around 11 billion dollars. By controlling the flow of the river's waters, the dam may provide flood pro-

tection to 15 million people living downstream, and is expected to reduce annual flood damage in the region by an average of 100 million dollars. In addition, the higher water levels behind the dam will make the river navigable as far inland as Chongqing, which is several hundred miles upstream. Each year the station will generate about 84 billion kilowatt-hours of electricity.

China leads the world in small hydro development, with approximately 60,000 small plants now providing 13,250 megawatts of generating capacity. These facilities account for about 30 percent of China's total hydropower capacity and provide a sustainable source of electricity for rural communities with few or no other sources of power. The generating equipment is generally built with Chinese labor and materials, thereby providing a new source of employment and reducing dependence on imported technology. It is estimated that at least 60 factories, employing more than 8,000 people, are engaged in manufacturing components for small hydro plants in China. The World Bank's evaluation of 31 developing countries — which between them more than doubled their hydropower capacity during the early 1980s — found that at least 28 of the coun-

114-115. With more than 600,000 isolated villages, in which about 70 percent of India's 950 million people live, solar water pumps are destined to play an important role. The pictures show a woman in Tamil Nadu spreading her saris to dry after having washed them at a PV-operated water pump.

tries had established programs to develop their small hydro resources. Major developers of small hydro in industrialized countries include Sweden, with about 1,350 plants (8,400 megawatts); the United States, with 1,700 plants (3,420 megawatts); Italy, with 1,400 plants (1,969 megawatts); and France, with 1,500 plants (1,646 megawatts).

In Europe the pure philosophical concept of hydrology changed into scientific observation during the Renaissance, through the work of such luminaries as the Florentine artist and scientist Leonardo da Vinci. Since then new scientific concepts and knowledge have emerged as new tools for investigation, such as

computers and nuclear technology, are employed. One new concept is to treat the hydrological cycle as a dynamic sequential system, which consists of an input, an output, and the working medium water (known as throughput) passing through the system. For example, a drainage basin or watershed is a system in which the input is the rainfall and groundwater inflow; the output is the evaporation, infiltration, and runoff; and the throughput is the water moving through the watershed. Using this system concept, the hydrological cycle can be readily interpreted by modern system analysis techniques and can be simulated by mathematical models and computers.

116-117. *The use of water mills to meet power demands in the Netherlands has given way to PV systems that ensure a reliable power supply in remote locations, as is the case at Dalfsen weir, belonging to the Bezuiden de Vecht water board district.*

118. *These Rajasthani women in India are looking forward to the day when they will enjoy the luxury of using solar water pumps instead of lining up and using muscle power to draw water at the village hand-pump.*

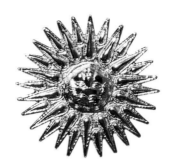

Blue Energy of the Ocean

In the Greek version of genesis, the Earth (Gaea) mates with Heaven (Uranus) and gives birth to the sea (Pontus). She then separates from the god of the sea (Posiedon) to become all other beings, including mountains. Variations of this belief, in which it was from the ocean that the Earth emerged, are common to traditions around the world. Mâui, with his renowned strength, pulls out of the deep ocean a giant ulua fish that has the power to transform itself into land. In a Vedic myth, Vishnu helps the weakened gods by churning the ocean for *amrita*, the "water of life," when they beseech the sun god for renewed energy and immortality. A Sumerian myth has it that when the gods send a deluge to "destroy the seed of humankind," Ziasudra (the counterpart of the biblical Noah) prostrates himself before the sun god, Utu, who then comes forth out of the sea to illuminate and warm up the Earth, and thus creates the "divine paradise-land where the sun rises."

We now know scientifically that life originated in the ocean, the great masses of water that cover 71 percent of the Earth's surface. They are a vast storehouse of solar energy that creates the ocean currents and the winds that produce waves on the ocean surface by solar heating and cooling, precipitation, and evaporation. The amount of solar energy that the tropical oceans absorb every week is estimated to be roughly equal to the energy content of the world's entire oil reserves — about 1 trillion barrels of oil. The total amount of power in waves breaking along the shorelines worldwide is said to be between 2 and 3 million megawatts, which is roughly equal to the generating capacity of 3,000 large power stations. The rising and falling tides dissipate an amount of energy roughly equal to that of the waves, and the stored solar energy is a potential resource for the operation of heat engines that can convert the ocean's thermal energy into electrical energy. Several different types of wave energy devices have been designed to extract the ocean's energy, which is immense but very difficult to harness as it is spread over millions of square miles of ocean surface and tens of thousands of miles of coastline.

Wave Power: Ocean waves are created by the winds blowing over the surface of the water. Their size and energy content depends on the speed and duration of the wind and its fetch — the distance on the ocean surface over which the wind blows. Although Masuda in Japan has been working on the extraction of ocean energy since 1945, research into at least ten different ideas for converting ocean energy to electricity began in several countries in the 1970s. With intense programs of model tests and theoretical work from brilliant mathematicians, notably in Norway and Great Britain, the size of full-scale plants was reduced and high

119. *In the Tao creation myth, Fuxi and Nu-wa represent the sun and the moon as they emerge out of the ocean in a semi-reptile form. Symbolizing creativity and renewal, Fuxi holds a carpenter's square, Nu-wa a pair of compasses. 7th-century painting on hemp from Turfan, China. Courtesy: National Museum of Korea, Seoul.*

120. *A 3rd-century B.C. Greek icon in gold of the sun god Helios. Courtesy: Louvre Museum, Paris.*

121. *(following pages) An aerial view of the two wave power stations that started operating in Norway on the island of Toftestallen in late 1985. The power station built by Kvaerner was based on the oscillating water column (OWC) design, while Norwave's Tapered Channel plant is known as the Tapchan.*

conversion efficiencies extended to longer and longer waves. Special turbines and hydraulic machines for power conversion were invented and the first tentative full-scale designs were produced in 1982 when several British teams and other scientists elsewhere in the world believed they had made the breakthrough.

Four types of interface have been proposed, and among them the simplest wave devices are symmetrical with respect to the wave's direction. Water acts on interfaces by exerting pressure. Output can be doubled for unidirectional waves if they can be made asymmetrical so as not to transmit waves astern. The most popular is the water-to-air junction used in devices called oscillating water columns (OWC) developed by Masuda in Japan and since taken up by many others. Pressure can also be transmitted to air contained in a flexible bag, as first suggested by Michael French at the University of Lancaster and later developed by Sea Energy Associates for their "clam" designed at Lanchester Polytechnic in Coventry. Water pressure can be transmitted to moving, hard-skinned structures as in the raft designs of Sir Christopher Cockerell in Great Britain and Glen Hagen in America, the buoy of Budal and Falnes in Trondheim, the submerged cylinder of David Evans at Bristol, and Steven Salter's "duck" developed at Edinburgh University.

The Masuda project in Japan started as a box with its bottom open to the sea. Wave action caused water within the box to rise and fall so as to exert an alternating pressure on the air above it. The power of water motion is harnessed in a vertical pipe and directed through a turbine, generating electricity for lights or other equipment on the buoy. The Japanese Maritime Safety Agency began using wave-powered weather-monitoring and navigational buoys in the early 1960s, and since then hundreds of such 70-watt and 120-watt buoys have been installed, including many equipped with microprocessors to monitor wave pressure.

Japanese researchers, who have been testing ocean energy devices in the Sea of Japan since 1977, found that significant amounts of electricity generated at the Kaimei plant could be transmitted to land-based facilities. However, as the electrical power generated at Kaimei was expensive, more tests were conducted during 1985–86, when scientists tested several different types of turbine with the goal of enhancing wave-energy conversion efficiency and lowering generating costs. Since the mid-1980s, the Kaimei plant has served as a platform for testing the operation of wave-energy devices and electrical components in the marine environment. Thereafter, many more small experimental wave-energy conversion devices have been tested at various Japanese coastal locations. In 1983, researchers at the Japan Marine Science and Technology Center (JAMSTEC) in Yokosuka installed Japan's first OWC device on a rocky shoreline near Tsuruoka City on the northwest coast. It did not live up to the expectation of generating 40 kilowatts of electricity, only producing an average of 11 kilowatts. Since 1989, scientists from Japan's Port and Harbor Research Facility have been testing a 50-kilowatt OWC plant installed on a harbor breakwater at Sakata.

The Japanese Takenaka Corporation has built a full-scale wave-power generating plant incorporating the constant air pressure tank system at Katagai in Kujukuri town, off the coast of Tokyo. The facility consists of ten OWCs with a total generating capacity of about 30 kilowatts, which is sufficient to supply electricity to ten houses. The converters are installed in a row on the shore embankment facing the sea. They also serve to dissipate the wave energy that cannot be utilized. The valve mechanism controlling air currents is placed in a cylindrical structure with an underwater opening. This OWC device has several original features. The constant air pressure tank equalizes fluctuating air pressure transmitted by the air duct and provides a flow of air at constant pressure to power the turbine. It also has the capability to store pressurized air temporarily, and the fluctuations in the amount of

122. Holding a mirror (the sun symbol) in one hand and a reptile (representing water) in the other, this angel by Piero Pollaiuolo (1443–1496) was a favorite icon of Mediterranean sailors during the Renaissance. Courtesy: Uffizi Gallery, Florence.

123. The plant built by Kvaerner

consisted of a 60-foot-high (18-meter-high) steel cylinder in the shape of a "bottomless bottle" placed on the seashore. As the waves struck the "bottle," the continuously oscillating water column rising in the OWC plant created an air flow that drove a Wells turbine, generating up to 500 kilowatts of electricity.

122

air that flows in from the inlet ducts can be equalized by the air tank floating up and down, while the constant air flow can be transmitted to the turbine from the outlet duct. When the generated power (three-phase AC) is low because of light wave action, grid power is used to compensate for the shortage, and when the ocean power is excessive it is absorbed by a resistor.

The Japanese have designed and tested a small-scale model of another type of wave generator called the "Mighty Whale." This floating platform equipped with OWC generators and moored offshore is a prototype of a larger plant that would be able to extract up to 60 percent of the energy in the incoming waves and to reduce the average wave height by about 80 percent. A series of Mighty Whales installed along a shoreline would not only generate electricity, but would also provide sheltered areas suitable for aquaculture and for recreational activities, such as swimming and wind surfing.

The Indian Institute of Technology in Madras completed India's first wave-energy facility in 1991, near Trivandrum in the state of Kerala. The 1-million-dollar plant, based on an OWC design, has a maximum generating capacity of 150 kilowatts. Indian researchers expect that the facility will generate an average of about 25 kilowatts during the months of December through March, a period when the sea is relatively calm, and about 75 kilowatts from April through November when the sea is rougher. The facility's maximum power generation is anticipated to be during the summer monsoon season, when its output may average 120 kilowatts. The Kerala state government has identified about ten other promising sites for wave-energy projects and has already proposed the construction of a 2-megawatt plant near Quilon in southern India.

The first two prototype wave-power stations in Europe began operating in late 1985 on the island of Toftestallen, about 18 miles (30 kilometers) northwest of Bergen in Norway. One of these power stations, built by Kvaerner, was based on the OWC design and consisted of a 60-foot-high (18-meter-high) steel cylinder in the shape of a "bottomless bottle" placed on the seashore so that its lower end extended beneath the water surface. The efficiency of oscillating water columns had been increased by changing the shape of the underwater structure to incor-

---

124. *(previous pages) A series of Kvaerner plants installed along the seashore could produce a great deal of electricity, as shown in this artist's conception.*

125. *The Tapchan is a funnel-shaped channel blasted out of the rock through which waves are funneled from an*

opening to the sea. As the walls narrow, the wave height increases until the water reaches the top of a wall and flows over it into a storage lagoon. The water then flows out of the lagoon and back to the sea through a 350-kilowatt Kaplan turbine.

porate a side opening and by the addition of "harbors" — a Norwegian innovation to induce a secondary resonance. As the waves struck the "bottle," the continuously oscillating water column rising in the OWC created an air flow that drove a Wells turbine, generating up to 500 kilowatts of electricity. The OWC performed well for four years until the 30-foot-high (10-meter-high) waves of a severe storm destroyed the plant in December 1988 — an unfortunate occurrence that nevertheless illustrates the tremendous power of wave energy.

Norwave's plant, known as the Tapchan, is a device in which the interface does not move. Designed and built in 1985 at Toftestallen by Even Mehlum, it is a tapered-channel system with a generating capacity of about 350 kilowatts. The interface is a funnel-shaped channel blasted out of the rock to a carefully calculated shape, with dimensions suitable for resonance with the local wave spectrum. The waves are funneled through a horn-shaped collector with an opening to the sea of about 2,000 feet (600 meters) into a 300-foot-long (92-meter-long) concrete channel that tapers from about 10 feet (3 meters) wide at the entrance to less than a

foot (one-third of a meter) at the far end. The tapering channel is cut 23 feet (7 meters) below sea level and rises up to about 10 feet (3 meters) above. As the walls narrow, the wave height increases until the water rises into an elevated storage lagoon with an area of 86,000 square feet (8,000 square meters). The water then flows out of the lagoon back to the sea through a 350-kilowatt Kaplan turbine.

The idea is related to the pioneering work of Noel Bott in Mauritius in the 1950s. He proposed a barrier to be built on the offshore coral reef of the island with a storage reservoir behind. By using this system, water could be stored and then drained back to the sea through a hydroelectric turbine. In 1995, Norwave and SINTEF in cooperation with the Norwegian-based INDONOR A.S. signed an agreement with the Indonesian Agency for the Assessment and Application of Energy (BPPT) to design and construct a 1.2-megawatt turnkey Tapchan project at Baron in Java. In the first tapered-channel prototype at Toftestallen, the rock blasting work was less accurate than it should have been. This deficiency will be improved in the Indonesian plant, on which work is cur-

126. Norwave's Tapered Channel is effectively a "water pump" in which wave energy lifts water up into an elevated reservoir. A plant based on this principle is being built at Baron in Java, Indonesia.

rently under way.

Research into so-called focusing devices was started in 1969 at SINTEF in Norway by Even Mehlum, and Norwave has since patented devices that, when submerged below the surface of the ocean, will act as focusing lenses to concentrate wave energy to where the power plant is situated. Focusing devices are not power stations, but rather passive installations that utilize ocean waves to concentrate their dispersed energy. They can be man-made barriers such as concrete caissons, or natural features such as reefs, which change a wave's direction and height so that its energy is concentrated into a small area, increasing its amplitude and power. At this focal point the waves are larger than in adjacent areas, and the water can be directed into a reservoir.

The European Union is encouraging wave-energy research based on OWCs, with projects in England, Ireland, and Portugal. The Portuguese are planning to install an OWC on the rocky coast of Pico in the Azores. Another system has been installed in a gully on the coast of Islay in the Hebrides. In 1991, this grid-connected 100-kilowatt wave-power device, which combines the resonant gully and the oscillating water column, became the first of its kind in Great Britain to become fully operational. Like the Norwegian OWC facility, the Islay wave-energy plant consists of an oscillating water column with a Wells turbine generator. A radically different type of wave-energy system — a ring-shaped clam of about 200 feet (61 meters) in diameter — has been designed in Great Britain by a group of researchers at Coventry Polytechnic and would be able to generate up to

127

2 megawatts of electricity. Circular clams are practical since they can be anchored to the sea bed off the coast to form a large power station, or a single unit can be used to provide energy to small off-shore islands or remote coastal localities not connected to a major power grid.

Surface followers are so called because they follow the contours of the waves. This French invention was first patented in 1799 in the shape of a buoyant "heaving float" that was attached to a ship and extended to the shore so that wave-induced motions of the ship caused the long lever to rise and fall; these oscillations were used to drive a pump or other equipment. The idea was that by linking two floating objects, or a floating object with a fixed platform, the energy of the moving water could either be transferred to a mechanical device or to a working fluid, such as water or air, and the captured energy in the working fluid could then be converted to electricity by a turbine. A novel "phase control" method of steering the movements of surface followers and OWCs for optimum power output was invented in Norway by Kjell Budal in the 1970s. These moving devices can capture wave energy from widths of ocean much larger than their physical dimensions.

However, it was not until 1974 that the design for the first high-efficiency surface follower was developed by Steven Salter and his colleagues. The Salter Duck — each of these cam-shaped elements looks somewhat like the head and beak of a duck — is threaded in a series onto a stiff shaft or spine and positioned so that the shaft is parallel to the incoming waves. The "beaks" of the ducks face directly into the waves

127. With ample circulation of oxygen-rich water, the Tapered Channel reservoir could be used for fish farming and serve as a recreation facility as well.

128. Focusing devices are passive installations that utilize ocean waves to concentrate their dispersed energy. The picture shows a focusing experiment in an outdoor test basin outside Oslo.

so that the force of the incoming waves causes the "ducks" to pivot about the spine — and this pivoting motion drives a hydraulic pumping system, which in turn powers an electric generator. In favorable conditions with high waves, Salter Ducks are said to capture more than 80 percent of the incoming wave energy.

A series of offshore wave-energy devices installed in a continuous line parallel to the coast would tend to reduce the energy and size of waves reaching nearby shorelines, and thus affect the natural environment of marine life since most fish and plankton remain in relatively shallow zones near the coast. Wave-focusing devices tend to increase erosion in areas where the magnified waves reach the shore, and increase sedimentation in adjacent areas where the wave heights are reduced. On the other hand, the reservoirs created by Tapchan systems can also be utilized as fish farms. The reservoir contains fresh, oxygen-rich water in a protected, controllable environment. Also, when demand is low, the excess power production can be used for producing ice and freezing fish. A prefeasibility study for the Azores indicated that the value of the fish farmed in a reservoir would actually exceed the cost of producing electricity. Many countries are now studying the impact of ocean energy devices on marine life, as a number of factors influence the distribution of small organisms and fish in the sea and in seashore reservoirs.

Tidal Power: The Egyptians recognized a connection between tides and tidal bores and the lunar cycle as early as 2000 B.C., and the Chinese scholar Wang Ch'ung writing in the 1st century A.D. declared that: "the rise of the wave follows the waxing and waning moon, and the natural processes of Heaven and Earth have remained the same since the most ancient of times. When the rivers fall into the sea there is nothing but rapid motion, but when the sea enters the rivers, the waters begin to roar and foam, doubtless because their channels are small, shallow, and narrow ..."

Tides are complex phenomena that not only reflect the influence of the sun and the moon in accordance with the laws of gravity, but also the rotation of the Earth about its axis and the motion of oceanic masses. Tides have been used as an energy source since at least the

129

11th century, when tidal-powered mills were built along the coasts of present-day England, France, and Spain. As recently as the mid-19th century, tidal mills were common in parts of Europe, Asia, and North America. However, the use of tidal energy had virtually disappeared by the early 20th century as hydropower and fossil fuels provided an inexpensive source of energy. Rapid growth in energy demand, combined with advances in construction methods and power-generating techniques, sparked renewed interest in tidal energy by the 1950s and 1960s.

A large 240-megawatt tidal power station to produce electricity was built between 1961 and 1966 on the site of the beautiful La Rance River Estuary in northwestern France. At this location the tide ebbs and flows into the estuary twice a day at a maximum rate of 635,000 cubic feet (18,000 cubic meters) per second, and the equinox tides have reached a record of 44 feet (13.5 meters) in amplitude. The La Rance dam is a half-mile-long (750-meter-long) structure built on a granite foundation that descends 42 feet (13 meters) below sea level. This has created a 6.5-billion-cubic-foot (184-million-cubic-meter) reservoir extending upstream nearly 12 miles (20 kilometers). It includes a lock on the left bank to allow boats to pass, and the station has 24 "bulb" generators, a permanent causeway, and six gates allowing the estuary to be filled and emptied quickly.

The operational principle used at the La Rance tidal plant and in other parts of the world is the same as that of the traditional tidal mill in which power was generated through a tidal damming system equipped with sluice gates and a paddle wheel. The French innovation of "bulb" units permits the turbines to be used in both water-flow directions so that power is pro-

129. A full-scale wave power generating plant incorporating a constant air pressure tank system was designed and built in Chiba, Japan. Its maximum output is 30 kilowatts, which is sufficient to supply electric power to about ten houses. A flounder farm using an underwater pump and blower is driven by its supplementary power.

duced during filling as well as during empty-ing of the basin — a double-action cycle. At this plant, a total of 24 turbines each with a capacity of 10 megawatts generate up to 600 million kilowatt-hours annually, or 90 percent of the power requirements for the 300,000 inhabitants of the city of Rennes. A 20-kilowatt cathodic protection system uses electric current to pro-tect the turbines from the corrosive effects of seawater. The success of this ocean energy pro-ject and the new eco-balance with its large and diversified number of species now living in and around the La Rance Estuary encouraged Electricité de France to propose the construc-tion of a 30,000-megawatt tidal energy project by closing the bay of Mont Saint Michel. But unfortunately, the French authorities diverted the available funds towards the construction of environmentally hazardous nuclear plants.

North America's first tidal power plant

began producing electricity during August 1984. The 17.8-megawatt plant is located on the Annapolis River about 100 miles (160 kilome-ters) west of Halifax, Nova Scotia, and pro-duces about 30 to 35 million kilowatt-hours of electricity annually. When the tide rises at the Annapolis facility, seawater flows through open sluice gates into the reservoir behind a 750-foot-long (230-meter-long) rock dam. When the tide begins to fall, the gates are closed so that the water level in the reservoir remains higher than that outside the dam. When the water is at least 5 feet (1.5 meters) higher in the reservoir, the gates are opened and the generators begin pro-ducing electricity. The plant reaches its maxi-mum output when the initial difference between the water levels is 18 feet (5.5 meters).

Canadian utilities have considered build-ing much larger tidal plants in bays and basins adjoining the Bay of Fundy, which has the high-

130. This OWC system, which converts wave energy to air pressure, is fitted with converters that are installed in a group on a shore embankment, bank protector, or seawall. The converters also function as a wave dissipation facility to diminish wave energy that cannot be utilized.

est tidal range in the world with fluctuations of up to 50 feet (16.5 meters). In the late 1970s, the Bay of Fundy Tidal Power Review Board conducted an assessment of more than 20 potential tidal power sites in Nova Scotia and New Brunswick, and identified three locations with the greatest potential for development. One of the most promising sites — the Cobequid Bay/Minas Basin of Nova Scotia — could provide about 3,800 megawatts of generating capacity, according to the review board.

However, no new tidal plants are currently planned in Canada because of the high projected costs of building a new facility and the availability of alternative energy sources, including untapped hydropower resources.

The use of tidal energy in the People's Republic of China began in 1956 with the installation of a 40-kilowatt station in Shashan, and since then eight tidal power stations have been established with a total generating capacity of 6,210 kilowatts. China's first two-way tidal

131. The aerial view of Maremotrice on La Rance River Estuary shows the world's first tidal power station to produce electricity. It was built in France between 1961 and 1966 and is based on the ancient Saint-Suliac tidal mill on the same spot.

plant — capable of generating electricity during both the rising and falling tide — began operating in 1980 on the Jiangxia Creek. The 4.1-million-dollar facility had a generating capacity of 500 kilowatts when first built, and its capacity was expanded to 3.2 megawatts during the spring of 1986. With this increased capacity, the plant is capable of producing 10 million kilowatt-hours per year.

In the former Soviet Union, the relatively small Kislogubsk Tidal Power Station on the White Sea has been operating since August 1969. Unlike the La Rance facility, the 400-kilowatt Kislogubsk project was built with floating construction techniques, not with cofferdams — watertight enclosures installed or built under water and pumped dry to allow construction within. Cofferdams are one of the most expensive components in constructing a tidal facility, costing 20 percent of the total spent on the La

Rance project. Researchers have also conducted preliminary studies for what would be the world's largest tidal facility — the 15,000-megawatt Mezenskaya Tidal Power Plant in the estuary where the Kuloy River flows into the White Sea. The tidal fluctuations here average nearly 30 feet (9 meters). The Mezenskaya power plant would have a barrage about 58 miles (93 kilometers) long, enclosing about 540 square miles (1,400 square kilometers) of sea.

The British government is considering an 8,600-megawatt tidal facility on the Severn River Estuary in southwestern England. The Severn plant calls for the construction of a 10-mile-long (16-kilometer-long) dam, containing 166 massive concrete sluices and 216 turbine generators. A series of locks would be built into the barrage to allow even large tankers to pass safely into and out of the estuary. The plant would supply an estimated 17 billion kilowatt-

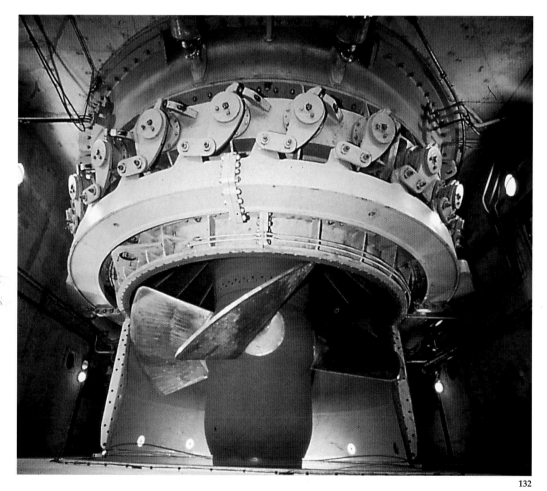

132

133

132. La Rance Dam includes a boat lock on the left bank, a station with 24 "bulb generators," a permanent causeway, and six gates allowing the estuary to be filled and emptied quickly.

133. La Rance tidal power station is composed of a turbine and an alternator generating 10,000 kilowatts of electricity. The specially designed "bulb generators" function whether the tide is rising or falling and can also work as pumping engines.

hours of electricity — the equivalent of 7 percent of Britain's total electricity demand in 1991. Another large tidal power plant has also been proposed in the Mersey Estuary near Liverpool, northwest England. According to one proposal, the Mersey barrage would be about 1.1 miles (1.8 kilometers) long and would contain 27 turbine generators with a total capacity of 621 megawatts. The facility would generate about 1.2 billion kilowatt-hours of electricity per year.

Although rising and falling tides dissipate energy at the rate of about 2 to 3 million megawatts, only a tiny fraction of this total is actually usable. Tidal power plants are practical only in areas with a relatively high tidal range — about 16 feet (5 meters) or more. It is estimated that only about 23,000 megawatts of continuous electrical power could be generated from tidal power worldwide — about 1 percent of the electricity that could be produced by hydropower.

Hence it will be long time before the ocean's tides become a major global energy resource.

Ocean Thermal Power: Jacques Arsene d'Arsonval, a French physicist and engineer, was the first to propose in 1881 that the temperature differences in seawater could be used to generate power. He suggested a "closed cycle" system in which a pressurized fluid such as ammonia is vaporized with warm surface ocean water to run a turbine generator to produce electricity. The vapor is then condensed with cold ocean water to begin the cycle again. A former student of d'Arsonval's, Georges Claude — the inventor of the neon light — followed up the idea and in 1930 built and tested the world's first "open cycle" system, which was located at the deep-water Matanzas Bay on the northwestern coast of Cuba. In this system, warm water itself is used as the working fluid. Since then, a third, hybrid system has been

134

135

134. The technology for generating electricity from the difference in ocean temperatures is known as OTEC — "ocean thermal energy conversion." This is an artist's rendition of the proposed 200-megawatt OTEC methanol plant-ship.

135. In 1983, the first experimental OTEC plant was established on an artificial island off the coast of Oahu in Hawaii.

136

designed that combines elements of the open-cycle and closed-cycle systems to produce both electricity and fresh water.

Known as Ocean Thermal Energy Conversion (OTEC), it is a system in which the solar energy absorbed by the ocean can be captured in regions where the temperature difference between the surface and deeper water is at least 36°F (20°C). The sun's heat warms the surface waters in the tropics to temperatures in the range of 80°–88°F (27°–31°C), temperatures that are maintained night and day throughout the year. But from a depth of about half a mile (1 kilometer) to the ocean's floor at an average depth of 2.5 miles (4 kilometers), the water temperature is only a few degrees above freezing. Thus the tropical oceans can be regarded as consisting of two vast reservoirs of water, one at a temperature of 80°–88°F (27°–31°C) and the other at 39°–41°F (4°–5°C). The electric-

ity generated by an OTEC system can be either transmitted to the power grid or used at the facility itself to produce hydrogen through the electrolysis of water or to make other valuable products, such as ammonia and methanol, which can later be transported by pipeline or tanker to shore.

OTEC plants could also be used to provide a number of other products not available from any other renewable energy resource. Aquatic plants, such as seaweed, could be cultivated, harvested, and either burned directly or converted into liquid fuels. The cold, deep water used in OTEC plants contains fewer pathogens than surface water and has a much higher concentration of nutrients such as nitrates and phosphates. These relatively unpolluted, nutrient-rich waters could support mariculture operations for growing salmon, trout, abalone, lobster, and edible seaweed. The cold seawater could also be used

136. Research in both closed-cycle and open-cycle OTEC systems continues at the shore-based Keahole Point plant, which has already set the world record for OTEC power production at 255 kilowatts (gross).

137. An artist's rendition of an OTEC plant of the future.

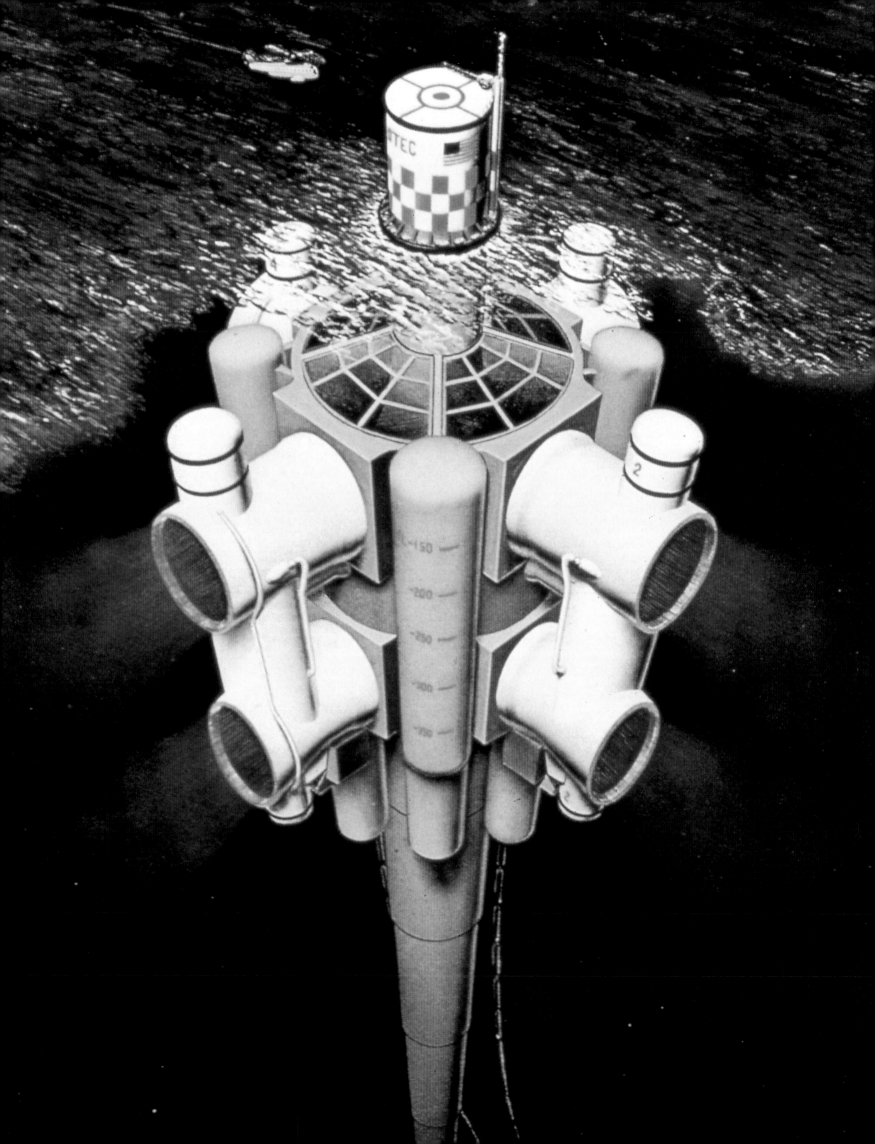

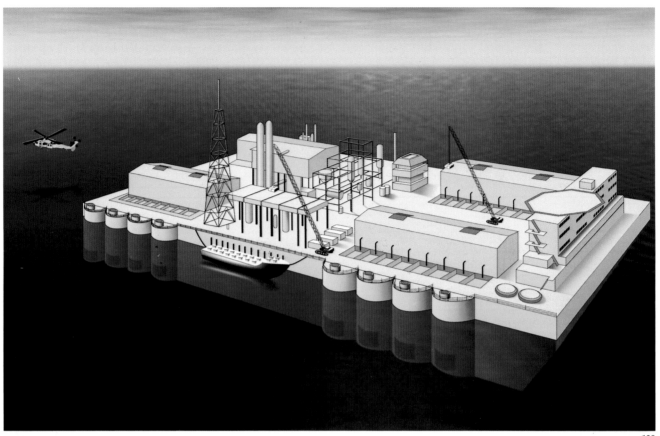

to air-condition nearby buildings and to maintain relatively cool soil temperatures in greenhouses, which would allow the cultivation of cool-weather crops such as lettuce and strawberries even in the tropics. In addition, the cold water could be used to condense moisture from the humid tropical air, thereby providing fresh water for crop irrigation.

As oil prices rose and technology improved during the 1970s, France, Germany, Japan, Sweden, the Netherlands, and the United States became increasingly interested in OTEC power generation. In 1979, the state of Hawaii and a group of American companies funded a closed-cycle mini-OTEC system mounted on a Navy barge — refitted with a power plant and a cold water pipe — moored in water 3,000 feet (1,000 meters) deep about 1.5 miles (2.5 kilometers) off the coast of Keahole Point, Hawaii. The floating 50-kilowatt (gross) plant operated for six months and proved a success, producing 10 to 15 kilowatts of net

power besides yielding valuable operational data in late 1979.

Encouraged by the mini-OTEC, the United States Department of Energy (USDOE) proposed to build a larger, 1-megawatt system and deployed a Navy tanker, S.S. *Ocean Sea Converter,* in 5,000 feet (1,500 meters) of water northwest of Keahole Point. This ship tested a 2,000-foot-long, 3-foot-diameter (650-meters-long, 1-meter-diameter) polyethylene cold water pipe attached to a heat exchanger, as well as other equipment for a working fluid system and cleaning techniques for biofouling and corrosion. The idea was to eventually construct a 40-megawatt closed-cycle offshore pilot plant, and the USDOE announced in 1982 its support for two closed-cycle plants — one mounted offshore and the other shore-based — at Kahe Point, Oahu. Unfortunately, ocean energy suffered a setback when oil prices dropped in the mid-1980s, and the project was dropped. Instead, a small land-based open-cycle experi-

138. In a futuristic scenario, some energy researchers speculate that OTEC plants could become the sites of the world's first floating cities, since they could produce power, air conditioning, and fresh water for inhabitants, as well as nutrient-rich cold waters to support the culture of fish, shellfish, and seaweed.

mental OTEC facility was installed at the Keahole Natural Energy Laboratory of Hawaii (NELH) site. It produces a net power of 40 kilowatts after subtracting the power needs for water pumping and other parasitic loads.

In 1981, a group of Japanese companies built a closed-cycle OTEC demonstration facility on the island of Nauru in the equatorial Pacific Ocean. In the plant, water at 86°F (30°C) from the ocean surface was used to evaporate Freon as the working liquid, and the resulting vapor drove a turbine generator. The vapor was then condensed by cold water at 45°F (7°C) pumped from a depth of about 200 feet (58 meters). The liquefied Freon was returned to a heat exchanger to continue the cycle. Although the plant produced up to 100 kilowatts of electricity during a one-year demonstration period, its net power output was no more than 35 kilowatts.

OTEC plants may be a particularly attractive power option in areas with limited supplies of both fuel and fresh water. A 10-megawatt plant could provide enough electricity to meet the needs of 10,000 to 20,000 people in industrial countries, and up to 100,000 people in developing countries where per-capita energy use is much lower. The plant could also produce up to 10 million gallons (40 million liters) of fresh water per day, sufficient for about 100,000 people in developing countries, according to Luis Vega, OTEC project manager at the Keahole Point NELH site. Thus, in many of the developing island nations with a population below 100,000, a single 1-megawatt to 10-megawatt OTEC hybrid facility might meet virtually all of the electricity and water needs.

In a surrealistic scenario, some energy researchers speculate that OTEC plants could become the sites of the world's first floating cities, since they could produce power, air conditioning, and fresh water for inhabitants, as well as nutrient-rich cold waters to support the mariculture of fish, shellfish, and seaweed.

Excess electricity could either be cabled to shore or used to produce fuels and chemicals for export to land. The pleasant climate of the tropical oceans might also make the floating cities attractive destinations for vacationers, who would provide tourist revenue for the OTEC-islanders.

Solar Ponds: Differences in the salinity of seawater and fresh water could also be used to produce power, although this area of ocean energy research is still at a very early stage of development. Arnold Goldman, the inventor of word processing, launched "solar trough" technology in 1979. He used this to build solar ponds that take advantage of a gradient in salt concentration — and therefore density — from the water's surface to the bottom of the pond, thus creating temperature differences between warm water at the deeper layers and cooler surface water. This drives a heat engine to generate electricity. In Israel, Harry Tabor constructed a pond with an area of 2.7 million square feet (250,000 square meters) at Beit Ha'Aravah near the Dead Sea and used it to power a 5-megawatt turbine of special design. Unlike other solar systems, the salt-laden solar pond can be used as a heat source day or night on a continuous basis once the temperature of its floor reaches the maximum of close to 212°F (100°C). Smaller systems were also built in Australia and California, although this kind of technology has yet to prove itself on economic criteria.

139

139. *Sunrise at Shimoda Beach. Wood-block print by Fuyo. Courtesy: Elizabeth Schultz Collection, Lawrence, Kansas.*

Orange Energy of the Geothermal

The original Hawaiians believed that the active volcano Kilauea, on the island of Hawaii, is home to a family of fire deities headed by the goddess Pele. She was driven from her original home in the western Pacific by her sister Na-maka-o-ka-hai, goddess of the sea, who persistently attacked Pele's home with floods and high tides. So Pele fled and made several attempts to build a new home on different islands of Hawaii. But each time her efforts were in vain as the water pursued her everywhere, quenching the fires in great explosions of steam. Finally, Pele found a place on Kilauea, where she lives safely in an enduring fire pit. Like her sister, Pele too is temperamental, sometimes wrecking havoc with earthquakes when angry, and on other occasions providing the health-giving geothermal water when her relationship is better with Na-maka-o-ka-hai.

Most ancient cultures worldwide revere volcanoes — such as Fujiyama in Japan — and the hot water from geothermic springs has been used for curing disease since time immemorial. The first emperor of China, Qin-Shi-huang (221–206 B.C.), is said to have discovered the "magical powers" of a Fairy Spring on Lishan Mountain near Xi'an that could cure all kinds of ills. This geothermal spring is so called because a fairy is said to have emerged from the water and demanded an apology from the emperor for not having made the customary animal sacrifice before entering the pool. In 747 A.D., another hot spring, in the Emperor Tang-Xuan-zong's palace, is described in a comment by the poet Li Bai: "...the hot spring water is so refreshing and pure that it washes away even the make-up of the Emperor's concubine Yang."

The geophysical features of the Hawaiian chain of mountains are typical of the nature of volcanoes generally. They were formed by volcanic processes operating over the past many million years — processes that are continuing even today. Rock and water in the Earth's crust are continually heated by the decay of radioactive elements and the intrusion of molten rock, or magma, from the Earth's mantle into the crust.

This stored heat, known as geothermal energy, can be extracted and used to warm buildings, generate

142

---

140. A volcanic eruption such as this in the Pacific Ocean represents a huge energy resource, though difficult to harness. Commercial utilization has been undertaken in some 40 countries for more than three decades, both for electricity generation and direct uses, such as space heating, horticulture, fish farming, and industry.

141. An American fire brigade symbol (1710).

142. Salt extraction from geothermal water in Zigong, Sichuan Province, China, during the Han Dynasty (206 B.C.–A.D. 220).

electricity, or perform other valuable functions. These resources exist in four major forms: hydrothermal systems, geopressurized systems, hot dry rock, and magma. Geothermal resources are generally most accessible in regions where the Earth's crust is relatively thin or has been disrupted by mountain-building (tectonism) and geologically recent volcanic activity — that is, activity within the past 10 million years. The amount of geothermal energy stored below the Earth's surface is enormous and represents 35 billion times the world's present total annual energy consumption. In reality, however, only a tiny fraction of this natural heat can be extracted from the Earth's crust.

Some 2,000 years ago, the Japanese and Romans began using low-temperature geothermally heated spring water for bathing and cooking. Later, during the Middle Ages, rudimentary geothermal district heating systems were built in Europe. In 1913, a geothermal power plant using a 250-kilowatt steam turbine started operating in Larderello, Italy — the deep wells produce superheated steam at great pressure, and the steam not only drives turbines to generate electricity but in addition contains chemical impurities that are recovered as valuable by-products. Today, geothermal projects are producing electricity in more than 20 countries, and the total world electrical generating capacity from geothermal resources increased from about 5,800 megawatts in 1990 to nearly 6,800 megawatts by 1995. Seven countries — the Philippines, Mexico, Italy, New Zealand, Japan, Indonesia, and El Salvador — each have more than 100 megawatts of geothermal electrical capacity. Other countries with geothermal resources include Nicaragua, Kenya, and Costa Rica, while countries such as Djibouti and Saint Lucia have the potential to produce all their electricity from geothermal resources.

Hawaiian interest in geothermal power dates back to September 1881 when the king of Hawaii, David Kalakaua, and his attorney general met with Thomas Edison to discuss the possibility of using steam from Hawaii's volca-

143

144

143. It is estimated that about 10 percent of the world's geothermal energy is used in Japan, although at present the total capacity of installed geothermal plants is only about 380,000 kilowatts. The Matsukawa plant has a geothermal power output of 23,500 kilowatts.

144. There are 14 geothermal power plants in Japan, of which 8 are public utilities (342,500 kilowatts) and 6 are private use (36,327 kilowatts).The geothermal power of the Kakkonda plant is 55,000 kilowatts.

noes to generate electricity for lighting Honolulu. Serious exploration started in the 1960s near the Kilauea Volcano's east rift zone in the Puna region of the island of Hawaii. The Hawaii Magma Power Company drilled four shallow wells in the area, but no exploitable resources were identified by these wells, the deepest of which was only 180 feet (56 meters). In 1972, months before the Arab oil embargo, the state legislature allocated 200,000 dollars to the Hawaii Geothermal Project to focus on the research capabilities of the University of

Hawaii for the identification and utilization of the state's geothermal resources. As a result, drilling began in late 1975 and was completed five months later, in April 1976, when a depth of 6,450 feet (1,966 meters) was reached. This well produced an average temperature of 675°F (357°C), a turbo generator was installed and this 3-megawatt plant began producing electricity in early 1982. However, after four exploratory wells failed to strike a commercially viable geothermal source in the 1980s, the three companies that were actively engaged in

145. *The Yamagawa geothermal power plant in Kagoshima has a geothermal capacity of 30,000 kilowatts. It is expected that by 2000, Japan may have developed 1 million kilowatts of geothermal power.*

147

the exploration of the geothermal power suspended their operation. So at present there is only one active 25-megawatt geothermal power project in Hawaii.

As a by-product of generating geothermal electrical power, a 4-acre site around this Hawaiian well was designated as the Puna Research Center for research, development, and commercialization of alternate uses of geothermal resources. The programs included such varied activities as dyeing fabrics, using geothermal hot water for aquaculture, and drying fruits and timber. Similar institutions

set up in other countries have found that geothermal water between 97°F and 115°F (36°C and 46°C) can be used to raise fish in aquaculture operations, while warmer water can be used to heat greenhouses, distill fresh water from seawater by evaporation, and dry organic materials such as fish, seaweed, and vegetables.

The United States is among the world leaders in geothermal energy development, with plant capacity totaling more than 2,816 megawatts. The technology used for both exploration and drilling, along with the various types of geothermal electrical cycles, is well doc-

148

146. The Blue Lagoon in Iceland is a favorite pool to which hundreds of people go to bathe in the geothermal brine from the Svartsengi power plant. The first outdoor pool for bathing in Iceland dates back to the 13th century; it was built of stone and clay by the famous saga writer Snorri Sturluson. Public utilization of geothermal water started in 1755.

147. The Pearl restaurant in Reykjavik rests on six huge storage tanks, each containing 1 million gallons (4

million liters) of geothermal hot water. The town (in the background) is heated entirely by volcanic hot springs. Other capital cities using geothermal water for heating are: Addis Ababa (Ethiopia), Beijing (China), Bucharest (Romania), Budapest (Hungary), Sofia (Bulgaria), Paris (France), and Tbilisi (Georgia).

148. Woman carrying geothermal hot water for private use. Stone sculpture by Asmundur Sveinsson Vatnsberinn, Reykjavik, Iceland.

umented. California's first geothermal plant with an electrical generating capacity of more than 1,000 megawatts was built in 1960, at Mayacamas Mountains Geysers, about 90 miles (145 kilometers) north of San Francisco. These facilities were expanded in the early 1970s, and at present geothermal heat is being used directly in about 150,000 individual installations at some 400 sites, besides supplying more than 110,000 geothermal heat pumps used for space heating.

Japan sits on an extensive volcanic formation with practically unlimited recoverable geothermal power resources, which will never be exhausted if used properly. At present there are ten commercial geothermal power plants in nine locations and an additional five private plants with a total generating capacity of 444 megawatts. Of the seven plants operating in Kyushu in the private sector, the Hatchobaru Geothermal Power Plant is the largest in Japan

149. The Great Geyser, a 200-foot-high (60-meter-high) fountain of geothermal hot water, makes a great impression as it springs out of the vast snow-covered landscape of Iceland.

with a total generating capacity of about 35 percent of the entire country's geothermal output. During the 1973 oil crisis, the development of geothermal power generation accelerated in many countries. In Japan the "Sunshine Project" was inaugurated — with an investment of 2.3 billion dollars for the development of geothermal energy. As a result Japan now has more than 3,300 megawatts of direct-use geothermal capacity, followed by China with 2,410 megawatts; Hungary with 1,280 megawatts; and the former Soviet Union with 1,000 megawatts.

In several countries, relatively low-temperature geothermally heated water is being used for space heating. The Reykjavik, Iceland, Municipal District Heating Service, the largest program of its kind in the world, has tapped geothermally heated water for space heating since 1930, and by 1990 about 85 percent of all residential buildings in the country were heat-

150. The Yamagawa geothermal power plant in Japan is one example of the great effort being made to protect the environment by attempting to design and harmonize the plant installations with the surrounding landscape.

151-152. Only a small fraction of the enormous geothermal power—illustrated by Etna in Italy and

other volcanoes — has so far been harnessed. It is estimated that annual geothermal utilization worldwide will grow by a factor of 45 for electricity generation and more than 100-fold for direct utilization, reaching an installed capacity of 8,900 megawatts-electric and 30,000 megawatts-thermal by the year 2000.

151

152

ed by geysers — a term of Icelandic origin. The heat is used for a variety of applications, including district heating, space heating, and heating for industry, greenhouses, and aquaculture. Similarly, in one former Soviet project at Mostovsky near the Black Sea, several thousand homes, 500 acres of fish pond, 55 acres of greenhouse, and a livestock farm are all heated geothermally. A large, innovative project on the island of Hokkaido in northern Japan uses hot water discharged from the Mori Geothermal Power Plant to heat 31 greenhouses that cover a total area of 15 acres. Geothermal district-heating projects funded by the EEC during the 1980s are now providing substantial amounts of energy in Ferrara, Italy, the Paris Basin and Bordeaux in France, and at Mons in Belgium. Worldwide use of heat from relatively low-tem-

perature geothermal resources, primarily for the heating of homes and greenhouses and for bathing, saves the equivalent of about 4.1 million tons of oil annually.

Geothermal energy is one of the cheapest sources of electricity, second in cost only to hydroelectric power. The cost of electricity from most liquid-dominated hydrothermal systems is higher than that from steam-dominated ones, and is currently competitive in only a limited number of locations. However, it is estimated that with additional research and development into advanced technologies to extract energy from hot dry rock and magma resources, geothermal energy could eventually provide vast amounts of usable energy at a cost competitive with conventional sources.

Hence geothermal energy will probably assume an increasingly important role in the 21st century.

153

153. The use of geothermal heat for fish and fruit drying is now practiced in a number of countries, including Hungary, Italy, Macedonia, Romania, Russia, China, Japan, Iceland, New Zealand, and the United States.

154. Like the Greeks, who worshipped Hephaestus, the Romans worshipped Vulcan, the god of fire, during the Volcanalia festival, when the heads of Roman families threw small fish into a fire. Like the Roman high priest, flamen, the Parsi (Zoroastrian) community in India have their own priest who tends fire — seen here at a Parsi Navjote child initiation ceremony in Bombay.

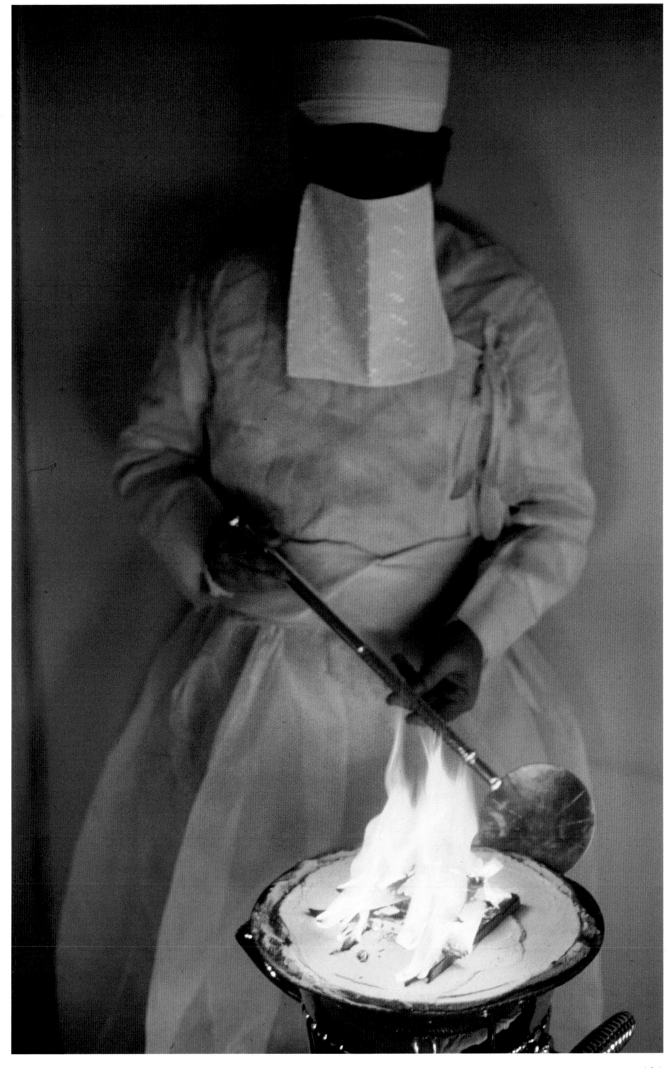

Pure Energy of Mirrors

The icon of Amaterasu in Shinto shrines in Japan is a round mirror that lures the sun goddess out of the dark cave, just as a mirror is still revered in some parts of southern India — for example, in the Sriman Narayan Swami temple at Swami Thoppu. A number of small disk mirrors excavated from the tomb at Shang-ts'un-ling in Honan (8th century B.C.) are believed to be the earliest bronze mirrors in China identified with the sun. Bronze mirrors widely used by the fire-worshipping Scythian peoples of the Eurasian steppe have been traced back to the Hui-style mirrors of pre-Han China, using the ancient practice of making mirrors of polished bronze incised on the nonreflecting side. The energy of the sun on the Honan mirrors is often shown by dragons interwoven with zoomorphs whose tails turn into volutes, symbolizing solar symbols. The Egyptian and Aegean bronze mirrors of pre-Roman Italy (10th–9th century B.C.) also show associations with the energy of the sun, as do Etruscan solar disks.

Ancient Chinese ceremonies are described in the book *Chou Li* (c. 20 A.D.), which reminds "the Sun-Fire priests of their duty to receive brilliant fire from the sun with a concave mirror ... in order to prepare brilliant torches for sacrifice." The environmental purity of solar energy derived from mirrors is invoked by the Greeks, who performed similar rituals in their Temple of the Vestal Virgins at Delphi. Plutarch tells how when barbarians sacked the temple and extinguished the sacred flame, it had to be immediately rekindled with the "pure and unpolluted flame from the sun" with the help of "concave vessels of brass" with which the holy women directed the sun's rays onto the "light and dry matter."

The Chinese, the Indians, the Greeks, and the Romans had all tried to concentrate the sun's rays onto an object with enough intensity to make it burst into flames. The "burning mirrors" of the Greeks were conceived as solar reflectors made of polished silver, copper, or brass — they probably consisted of many flat pieces of polished metal joined together to form a roughly curved surface. Later they were replaced by concave spherical mirrors made from single sheets of metal, and eventually by the first parabolic mirror, invented in the 3rd century B.C. by Dositheius, a Greek mathematician. A century later, in his treatise *On the Burning Mirror*, Diocles gave the first formal geometric proof of the focal properties of parabolic and spherical mirrors to create intense heat.

155. *While the water for tea is being heated by solar reflectors, a Chinese mother and daughter sit and chat as they enjoy the warmth of sunlight reflected by the solar water heater.*

156. *Korean bronze mirror representing the sun, c. 3rd-century B.C. Courtesy: National Museum of Korea, Seoul.*

157. *A mirror tempts the Japanese sun goddess Amaterasu to come out of her dark cave and save the world from extinction.*

157

158

159

160

The fable that Greeks had used burning mirrors to destroy the ships of invading Romans at Syracuse — attributed to Archimedes (212 B.C.) — seems to have fired the imagination of Ibn Al-Haitham, an 11th-century Arab scholar living in Cairo. He calls the burning mirrors "one of the noblest things conceived by geometricians of ancient times," and his mathematical elaboration of the concentrating powers of parabolic mirrors was later translated into Latin and circulated among several European universities during the middle of the 13th century. Then, around 1515, Leonardo da Vinci suggested that the mirror be used not for military, but for peaceful purposes. He proposed the building of a huge parabolic mirror that could supply heat for boilers in industry and warm up swimming pools. According to Leonardo, the sculptor Andrea del Verrocchio

employed a burning mirror to solder the sections of a copper ball lantern holder for the Santa Maria del Fiore Cathedral in Florence. Adam Lonicier, writing in 1561, records the technique of using spherical mirrors to heat submerged flowers to make perfume. During the latter part of the 16th and throughout the 17th century, almost every scientist investigated the mysterious powers of burning mirrors, though some of the accounts of their activities proved to be grossly exaggerated.

In the late 16th century, Villette, a French optician from Lyons, built several large spherical mirrors, the largest of which measured more than 3 feet (1 meter) in diameter. A report published in London claimed that the Villette mirror was able to make tin melt in 3 seconds, cast iron in 16 seconds, and cause a diamond to lose 87 percent of its weight. "In short," the article concluded,

158. An Indonesian artisan displays a decorative earthenware plaque embedded with a round mirror that would traditionally be installed on a rooftop to represent the sun.

159. A stylized bronze mirror worshipped by Namboodri brahmins in Kerala, south India.

160. An Indian priest meditates in front of a mirror representing the sun. "All are equal under the sun" is the motto of the Swami Thoppu Temple, which is exceptional in that any person can freely enter to worship, regardless of caste, creed, or religion.

161

"there is hardly any thing which is not destroyed by this fire".

In the 17th century, attempts were made to build mirrors that were both lighter and easier to handle, and a Dresden mechanic named Gartner constructed mirrors from wood, coating the concave inner surface with wax and pitch and then a layer of shiny gold leaf. The effort failed as the gold leaf quickly deteriorated, completely destroying the mirror's reflectivity. More successful was a German nobleman, the baron of Tschirnhaus, who hammered a single sheet of copper "scarce twice as thick as the back of an ordinary knife" into a mirror 5 feet (1.6 meters) in diameter, which could be easily handled by one man.

As larger mirrors could not be constructed with single pieces of metal, Peter Hoesen, an 18th-century mechanic, built his mirror from sections of hard wood covered with pieces of brass. Thus the German craftsman was able to build mirrors as large as 10 feet (3 meters) in diameter — three times the size of Villette's biggest reflector and almost twice as large as that of the baron's. These mirrors, by far the most powerful solar reflectors yet developed, concentrated the sun rays onto a target area around an inch (2 centimeters) in diameter, and a researcher who experimented with a Hoesen reflector 5 feet (1.5 meters) in diameter confirmed that "copper ore melted in one second, lead melted in the blink of an eye, asbestos changed to a yellowish-green glass after only three seconds, and slate became a black, glassy material in 12 seconds." Moreover, Hoesen's mirrors were easy to handle: "One can put it in any position using but one hand," remarked an observer.

It was against this background of ancient bronze mirrors in China and the subsequent Western technology based on Greek and Roman ideas in the 1st century that a French professor of mathematics, Augustin Mouchot, effectively constructed and used conical and parabolic dish collectors in the 1860s. During the 1980s, solar thermal electric generation —

the use of heat from solar energy to generate electricity — developed rapidly, as experimental facilities were tested in several countries. The results are now applied to modern systems such as the parabolic dish collectors in Solarplant One in the United States. Located on a 40-acre site near Warner Springs, California, the 4.9-megawatt system, built at a cost of 19 million dollars, collects sunlight with its 700 parabolic dishes, each containing an array of 24 reflectors 5 feet (1.5 meters) in diameter. Its field of collectors is divided into two sections and the mirrors are mounted on a lightweight, movable frame that tracks the sun throughout the day. The solar energy gathered by collectors in one section is used to vaporize water and then heat the resulting steam to a temperature of 530°F (275°C). Any water that does not evaporate during the initial heating is separated from the steam and returned to the collectors for additional heating. The steam is then piped into a second group of solar collectors, where it is further heated and used to run two turbine generators.

In another type of system, a "power tower" is surrounded by sun-tracking mirrors called heliostats, which reflect sunlight onto a receiver at the top of the tower. The world's first power tower began operating in Nio, Japan, on the island of Shikoku during early 1981 and was tested until 1986. In this plant, more than 800 heliostats with a total surface area of about 140,000 square feet (13,000 square meters) directed sunlight onto a receiver at the top of a 200-foot (61-meter) tower. The heat of the concentrated sunlight was used to produce high-temperature steam, which generated up to 1 megawatt of electricity. The plant was part of Japan's Sunshine Project, an on-going program to develop and test alternative energy technologies in that country.

A year later, the world's largest experimental solar power tower, the 10-megawatt plant called Solar One, was erected on a 130-acre site in Daggett, California, and operated

161. Mouchot's solar engine driving a printing press at the World Exhibition in Paris in 1878. The 15-foot (5-meter) diameter mirror provided enough energy to print a special newspaper entitled The Sun *at a rate of 500 copies per hour.*

during a six-year demonstration program between April 1982 and September 1988. Built in the Mojave Desert at a cost of 141 million dollars, the Solar One consists of a 300-foot (91-meter) tower surrounded by 1,818 mirrors with a total surface area of about 780,000 square feet (72,500 square meters). Its excellent heat retention quality was enhanced by pieces of rock stored in large insulated tanks of water — the heat transfer medium that was used to warm the caves of the Stone Age people. Solar One has now been refurbished as Solar Two, the central receiver of which is heated by concentrated sunlight to a temperature of about 1,000°F (565°C). This heat is transferred to molten nitrate salt, which is used to boil water and produce steam, which is passed through a

turbine to generate electricity. At present it is producing electricity for 10,000 homes in southern California during the day, and after sunset it can actually produce power for up to three hours due to its innovative energy collection and storage system.

Europe's first power tower, Eurelios, was sponsored by the European Economic Community and produced power in Adrano on the island of Sicily during a demonstration period from April 1981 to 1984. The plant's 180-foot (55-meter) central tower was surrounded by more than 180 heliostats, with a total collecting area of 67,000 square feet (6,225 square meters). Sunlight from the heliostats heated a receiver on top of the tower; this heat was used to boil water and generate superheated steam at 950°F

163

162. *The inventor, Cassini, presents his mirror to the French King Louis XIV, "Le Roi Soleil," in the Paris Observatory. With a diameter of 9 feet (2.7 meters), this high-quality burning mirror is one of the largest built at the end of the 17th century in France.*

163. *Different types of solar cookers are now available worldwide, but their use is limited, mainly because people are not aware of the availability of this new product. An Indian entrepreneur is trying to popularize a small-size solar cooker by commissioning beautiful Rajasthani dancers to display them in the open-air compound of a hotel in Jodhpur.*

(510°C), which was then harnessed to produce up to 1 megawatt of electricity. The facility also contained a heat storage system in which a mixture of salts called Hitec was heated by the steam to a temperature of about 800°F (425°C) and then pumped into an insulated storage tank. When electricity was needed during heavily overcast periods, the heat from the stored salt was used to produce steam for up to 30 minutes at a time.

Three other central receiver facilities were tested in Europe during the 1980s. Two of these — the 500-kilowatt International Energy Agency Central Receiver System (IEA-CRS) and the 1.2-megawatt CESA-1 plant — were built at Tabernas near Almeria in southeastern Spain. The

165

largest European system ever built, the 2.5-megawatt Themis, was located in Targasonne in the French Pyrenees and operated during a three-year test period between mid-1983 and mid-1986. Each of these experimental plants included a thermal storage system that allowed the facility to produce electricity for short periods when the sun was not shining.

At the Weizmann Institute's solar tower in Israel, installed in 1987, the 64 computer-controlled mirrors (heliostats) with curved surfaces reflect and focus sunlight onto as many as five independent target receivers. The solar field can provide up to 3,000 kilowatts of concentrated solar light, and methods are being devised to utilize the solar energy in chemical form.

4

166

167

164. *In accordance with the law in Israel that requires solar water heaters to be installed in all new apartment buildings up to eight stories high, about 900,000 systems are currently in use nationwide. This bird's-eye view is of Jerusalem's Me'ah Shearim quarter.*

165. *This drawing of "Cristo Sole" — after the 2nd-century mosaic of the sun god Helios, discovered in 1574*

beneath St. Peter's Basilica in the Vatican — was made by students at St. Xavier's School at Kalol in Gujrat, India.

166-167. *Cooking and water heating at St. Xavier's School is done entirely with solar collectors constructed by the students themselves.*

Research has shown that a "chemical heat pipe" based on the reversible reaction in which methane is converted into hydrogen and carbon monoxide is a feasible energy source. The gaseous reaction products are then cooled, stored at high pressure, and transported to a "methanator" plant in which hydrogen and carbon monoxide react to produce methane, releasing large amounts of heat that is used to generate steam, electric power, or both. The cycle is completed by returning the methane to the solar site. Another field of research at the Weizmann Institute is to focus solar heat directly onto a Stirling engine (also known as a heat engine) to produce electricity.

Researchers in the former Soviet Union have built and tested a prototype central receiver plant, the 5-megawatt SPS-5 near the village of Myosovoye in the Crimea Peninsula. At this facility, 1,600 mirrors with a total area of 430,000 square feet (40,000 square meters) focus sunlight onto a receiver atop a 300-foot (88-meter) tower. This concentrated solar heat is used to produce steam at 900°F (480°C), which drives electric turbine generators.

Unlike central receiver systems or power towers, in which light is concentrated onto a single central receiver, distributed solar thermal systems use a group of parabolic dishes or troughs to collect sunlight and concentrate it on a series of receiving surfaces. In distributed systems, each parabolic dish or trough works individually to concentrate sunlight. A parabolic dish concentrates light on a focal point, while a parabolic trough focuses light along a line. Pipes run between the focal points or along the lines where the sunlight is concentrated, and fluid pumped through these pipes is heated to very high temperatures by the intense light. The hot fluid may then be used to provide space heating in buildings, generate electricity, or produce heat or steam for industrial uses.

The world's largest solar thermal power projects in the early 1990s were the nine Solar

168

168. A Mancheng County Solar Energy School teacher instructs some of his 500 Chinese students in the art of solar water heating. Such units are used in the school building, in which about 90 photovoltaic lights are also installed.

Electric Generating System (SEGS) plants, located at three sites in California's Mojave Desert. The first project, a 13.8-megawatt unit called SEGS I, was completed in Daggett in December 1984. Built, designed, and manufactured by Luz International Ltd. of Los Angeles at a cost of about 62 million dollars, SEGS I consists of 560 parabolic troughs with a total reflective area of about 775,000 square feet (72,000 square meters). The parabolic troughs are motor-driven to track the sun as it moves through the sky.

The construction of a second plant, the 30-megawatt SEGS II, was completed in Daggett in October 1985. Five more 30-megawatt units,

SEGS III through VII, were built between 1986 and 1988 at Kramer Junction, California, about 40 miles (65 kilometers) west of the SEGS I and II site. During 1989 and 1990, Luz built its first two 80-megawatt units — SEGS VIII and SEGS IX — at Harper Lake, about 25 miles (40 kilometers) northwest of Daggett, bringing the total solar thermal capacity at all the SEGS plants to 354 megawatts. According to Solel (which took over the project from Luz after its demise), the cost of generating electricity from its 80-megawatt plants is less than that for nuclear power plants currently coming on line in the United States, even before the costs of nuclear

169. The Odeillo solar furnace, built in the Pyrenees Mountains in France in 1970, is a famous research device used for experiments on materials and processes at high temperature and high flux, equivalent to "ten thousand suns."

171

waste disposal and plant decommissioning are taken into account. The plants help both to meet peak demand and to reduce smog, because about 80 percent of the electricity from the plants is produced during times of peak power demand when air pollution in southern California is typically at its worst.

Other research teams were also developing solar troughs in the early 1990s, including an innovative design conceived at the University of Sydney. Tests revealed that improved collector surfaces, polar axis tracking, and an improved trough design can increase sunlight collection by nearly a quarter. Furthermore, by

adding vacuum insulation to the heat-carrying pipes it is estimated that the annual efficiency of the system may reach 20 percent, with peak efficiency of between 25 and 30 percent. The new design also has the ability to run for eight hours without sunlight by storing heat in an inexpensive bed of rocks — again recalling the Stone Age. In Israel another system to supply hot water and steam efficiently to hospitals, hotels, and industry is based on an optimized cascade of solar fields, with increasing levels of temperature generated through a combination of plastic solar collectors (preheating) and linear tracking mirrors with fuel or gas backup.

170. The 10-megawatt Solar Two tower in Barstow, California, has succeeded the Solar One system that had been converting solar thermal power into electricity since 1982. The 200-foot-high (61-meter-high) receiver is surrounded by a circular array of 1,818 sun-tracking glass mirrors (heliostats), each of which measures about 23 feet (7 meters) high. Heat is stored in large tanks of molten salt and then drawn off to generate electricity.

171. This 4.9-megawatt system, built on a 40-acre site near Warner Springs, California, collects sunlight

through seven hundred 30-foot (9.5-meter) parabolic dishes, each containing an array of 24 reflectors measuring 5 feet (1.5 meters) in diameter.

172. (following pages) The world's largest solar thermal complex is located in California's Mojave Desert. Its 1.5 million mirrors focus the sun's rays on tubes filled with synthetic oil with which water is boiled to produce steam in a heat exchanger. The steam is then used in a turbine to produce electricity.

Industrial Stone Age

The Ayarmacas in northern Chile and the Incas of Peru both believe that their hero Viracocha created humankind; but when they show disrespect to the deity, he turns them into painted stone dolls. Then, if they atone, he brings them back to life with the help of solar energy. In a Greek myth, only one man and one woman survived the great flood unleashed by the enraged Zeus, and that couple — Deucalion and Pirra — recreated humankind by throwing stones over their shoulders on Parnassus, the mountain favored by Apollo, the Greek sun god. The sun god, Utu, in a Sumerian myth, helps Dumuzi, the young husband of the deity Inanna, to escape from the netherworld by changing him into a stone. A Chinese myth has it that while the strongman Yü was digging canals to drain off floodwaters to the sea, he married a T'u girl, who gave birth to a son, Ch'i, in the form of a stone — like his father, Yü. The ammonite shell "stone" *salagrama* represents the Indian sun god Vishnu, and the stone *brahmanada* is Brahma's "golden egg"— called the "resplendent sun" in the *Rig Veda*.

The origin of modern science may also be traced to a stone since it was the accidental discovery of "lodestone" that launched the unprecedented scientific and industrial revolution of our time. The stone's remarkable magnetic properties were first noticed by a Frenchman, Petrus Peregrinus de Maricourt,

in the 13th century, and were subsequently studied in greater detail by William Gilbert of England. In 1600, he wrote a systematic treatise on the Earth's magnetism, identifying the force between two objects charged by friction and calling it electric — the energy that eventually illuminated Thomas Edison's lightbulbs in 1882.

At the same time, because of the accelerated use of energy, efforts were being made in Great Britain to replace animal and human muscle power with mechanical energy. A breakthrough finally occurred with the development of the steam engine. It provided a long-awaited solution to the difficulties encountered in coal mines — since the deeper one dug, the more difficult it became to drain the infiltrating underground water from the pits. At one coal mine, 500 horses were needed to raise the water, bucket by bucket. The solution was found in 1698 by Thomas Savery with his "fire engine," followed by Thomas Newcomen's simple and sturdy "atmospheric steam engine," which worked a piston up and down a cylinder. In 1784, James Watt succeeded in converting the back and forth movement of the piston into rotary motion so that steam could be used not only to pump water out of mines or operate furnaces, but also to drive the wheels of industry and transport — particularly the railway engine. Initially, locomotive traction was introduced for coal haulage in 1825 on the Stockton

'3 a-b

---

173. *The PV-generated solar energy that on July 4, 1997, enabled (A) the Mars Pathfinder spacecraft to perform the incredible feat of landing on the Martian surface, some 120 million miles (191.5 million kilometers) away from Earth, and (B) the Sojourner rover to sniff the rocks (including "Yogi" at right) with its Alpha Proton X-Ray Spectrometer, is yet another landmark in scientific research that started in the 13th century with the discovery of lodestone's magnetic properties and*

eventually led to the production of PV electricity.
Courtesy: Jet Propulsion Laboratory, California Institute of Technology, Pasadena, U.S.A.

174. *The sunshine in London on the eve of the Industrial Revolution inspired Nicolas Briot (1633) to make this gold medal.* Courtesy: Sotheby's Collection, New York.

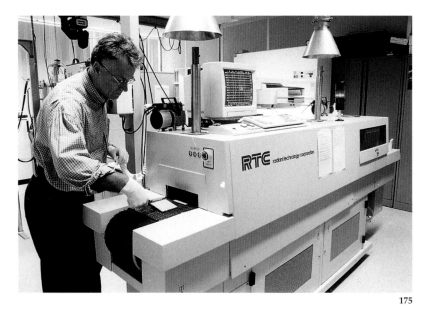

and Darlington Railway line, but to everyone's astonishment, passenger revenues exceeded freight receipts when the new Liverpool to Manchester Railway started in 1829. This was the beginning of the era of the railway, called the Iron Horse.

During the 17th and early 18th centuries a great deal of interest was shown in research on the electrostatic effects that lodestone's magnetism had revealed. Electrical conductors were discovered by Stephen Gray in England in 1729, and in 1733 Charles François de Cisternay Dufay, gardener to the king of France, announced the existence of two types of electric force: an attractive force, already known, and a repulsive force — to which Benjamin Franklin attributed the plus and minus signs, respectively. There was laughter when Franklin said that lightning was electric — until he proved it. Electrostatic effects attracted the attention of several other scientists, such as Sir Isaac Newton and Joseph Priestley, and the 18th century and early part of the 19th century saw substantial mathematical advances, mainly conducted by two physicists, Henry Cavendish and Charles Augustin de Coulomb, complemented by the work of the mathematician Siméon Denis Poisson.

Scientific curiosity was also directed to the source of it all, the sun, and its enormous output of solar energy. It was calculated that some

5.4 million exajoules of solar energy strike the Earth's upper atmosphere each year, one-third of which is reflected back into space, with another 18 percent absorbed by the atmosphere, much of it converted into wind. This leaves 2.5 million exajoules that reach the Earth's surface. Scientists also succeeded in showing that light and its invisible counterparts — radio waves, infrared, ultraviolet, and X-rays — are the purest forms of energy. These "electromagnetic radiations" are created by the movement of electrons, sometimes quite gently, as in the radio antenna, and sometimes violently, as when a beam of fast-moving electrons is suddenly halted by the target in an X-ray tube. The normal "jumps" of electrons in atoms are of intermediate intensity. All these radiant forms of energy can travel through empty space — from the sun to the Earth.

Thus research intensified to explore energy-conversion devices in which mechanical, chemical, or electrical energy could be changed from one form to another. Attention was focused particularly on fuel-burning heat engines, electric generators, batteries, photoelectric cells, and so on. Thomas Newcomen's engine, which converted less than 1 percent of thermal energy into work, was soon replaced by Watt's steam pumping engine. Watt's engine, invented in 1763, had an efficiency of more than 2 percent. By 1900 many improvements, including three separate expansion cycles and higher steam temperatures, had raised the efficiency of the steam engine to about 17 percent. The original steam engine has since been replaced by the more compact and efficient steam turbine. With improvements and the use of very high temperature steam, the efficiency of thermal energy conversion is now approaching about 40 percent.

The operation of a gas turbine power plant, such as a turbojet engine, is described by the Brayton cycle, originally devised for reciprocating engines — engines in which combustion takes place at constant pressure in the cylinder. The urgent need for improved aircraft propulsion during World War II led to the development of efficient, lightweight air compressors and gas turbines that made possible the turbojet engine operating on the Brayton cycle. Compressed air, heated by the combustion of kerosene, is used to

175. In the manufacture of silicon cells, a variety of ovens are used for screen printing. This is a conveyor belt oven containing various zones where strong infrared lamps are focused on the silicon wafers.

turn a turbine to drive the compressor, while excess energy accelerates the exhaust gas to high velocity for producing thrust.

While the science of electrostatics had been progressing and its fundamental laws were becoming clear and codified, an Italian professor of anatomy, Luigi Galvani, noticed that electrical sparks produced by a nearby machine caused muscular contraction in a dissected frog — a mystery whose explanation was discovered by the physicist Alessandro Volta, who in 1792 correctly conjectured that the effect was due to contact between the metal and the moist body. In 1800 Volta constructed an electrochemical cell for producing electricity from two metals in contact with a salt solution. A battery for producing higher potential, called Volta's pile, was made of alternate plates of silver and zinc separated by cloth or paper soaked in a salt solution. Many types of batteries were developed during the 19th century, and this source of electricity made possible early electrochemical discoveries. The lead-acid secondary battery for storing electrical energy was developed by Gaston Planté of France in 1859. Since then, new types of batteries have been developed, and recently research has accelerated, as storage of electricity is one of the main problems facing solar energy.

In 1820 a Danish physicist, Hans Christian Ørsted, announced his discovery that a magnetic compass needle was influenced by current in neighboring conductors, and by the end of the year a French physicist, André Marie Ampère, had shown that two parallel wires carrying current attracted or repelled each other depending on the direction of the current. As a result he formulated the mathematical laws that govern the interaction of electric currents with static magnetic fields. A number of other physicists conducted research into the subject: in 1827 Georg Simon Ohm published the law of the power of a material to conduct static electricity; Michael Faraday's discovery of electromagnetic induction was announced in 1831; and in 1841 James Prescott Joule published his theory on the heating effects accompanying the flow of electricity in conductors.

Other scientists in Europe, mainly in Germany, working on the mathematical theory of

176

electricity and magnetism included Franz Ernst Neumann, Wilhelm Eduard Weber, and H.F.E. Lenz — while Hermann von Helmholtz developed the theory of the relation between electricity and other forms of energy. As a result of research carried out by a number of scientists, the connection between electromagnetism and optics had also become apparent by 1860, as Scot James Clerk Maxwell's celebrated equations were to confirm. Maxwell's greatest distinction is that he finally synthesized electricity, magnetism, and optics into one coherent whole. It was in tune with Charles Darwin's first object lesson learned from his observations of a volcano during one of his famous voyages on board the *Beagle* — that "natural laws apply uniformly throughout time."

Exhaustive investigation of static energy-conversion devices — which use electrons as the working fluid in place of the vapor or gas in a dynamic heat engine — began because of their reliability and their lack of moving parts. In a thermoelectric generator, for example, electrons driven by thermal energy across a potential difference at the junction of two conductors made of dissimilar materials create an electric current. This effect was discovered in 1821 by Thomas Johann Seebeck of Germany when he observed that a compass needle near a circuit made of different conducting materials was deflected when one of the junctions was heated. He investigated

176. An overview of a typical production line of photovoltaic systems. While laminates are being prepared on the left, solar cell matrices are in production in the center, and the lamination process is being completed on the right.

many materials, including some now classified as semiconductors, which produce electrical energy with an efficiency of 3 percent. This efficiency was comparable to that of the steam engines of the time. Nevertheless, the usefulness of the discovery of the thermoelectric effect went unrecognized as a means for producing electricity because of Seebeck's interpretation of this as a magnetic effect caused by a difference in temperature. Only recently have thermoelectric generators come into their own, producing electricity in special situations in which static thermal-energy-conversion devices with high reliability but a modest efficiency are preferred.

Another static energy-conversion device that has been recently investigated extensively for the space program is the thermionic generator. Electrons are driven in a vacuum by thermal energy from a high-temperature electrode, across a potential difference, to a cooler electrode. Thomas Edison was the first to observe an electric current between an incandescent filament and a cold electrode in an evacuated tube, but it was not until 1958 that the active development of thermionic generators began. Although their efficiency has since been improved greatly, the lack of suitable materials for use at very high temperatures and problems in construction have limited their reliability.

At present, all theories of the electromagnetic properties of metals, insulators, and magnetic materials are formulated in terms of electrons. Until now, nothing has suggested that the color, quality, and chemical behavior of all familiar matter could be explained through electricity and magnetism. But that is in the nature of physics: the falling of an apple led scientists to speculate about the power that holds the planets on their courses; the passage of electric current through a gas has given a clue as to why grass is green and what the force is that holds a stone together. It so happens that the name of one of the most creative scientists in human history — Einstein — literally means "one stone."

177

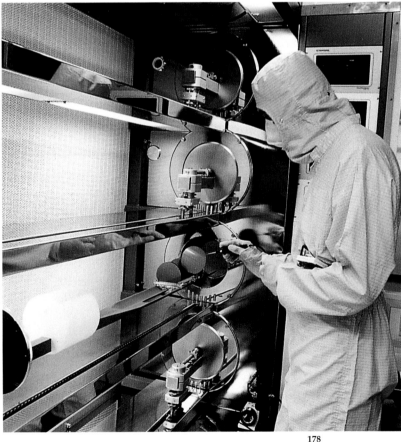

178

177. *Worldwide research is now being conducted in the field of thin-film silicon solar cell technology. The technique of high-temperature chemical vapor deposition (CVD) is used to manufacture high-grade polycrystalline silicon.*

178. *Amorphous silicon-based thin-film solar cells, and the alloys made with them, are manufactured in the two-chamber ultra-high-vacuum system known as AMOR.*

179

*179. A sun simulator is used to measure the
current/voltage characteristics of different types and
dimensions of solar cells as well as their spectral
response. Green light is focused onto the cell to
determine its spectral response while halogen lamps
provide the background lighting.*

142

Traditional Low Energy Architecture

Humankind, states the *Vedas*, was chiseled out of the orb of the sun by the heavenly architect Vishvakarma. And the tremendous power of the sun is suggested in the 4th century B.C. Chinese myth of Hsi-ho, "the mother of ten suns" — who lived in the branches of the Fu-sang tree in the Valley of Light. Taking the form of ravens, the suns took turns to appear in the sky on each day of the Chinese ten-day week. One day, through some confusion, they showed up all at once and would have reduced the Earth to ashes had not the famous archer Yi saved the world by shooting down nine of the ten suns, thus leaving only one sun in the sky. This Chinese myth has many parallels. Among the Shasta Indians of California, it is the coyote who slays nine of the ten brother-suns, and in the manner of the Chinese story, the national hero of the Golds in eastern Siberia shoots down two of the three suns when they begin to make the world unbearably hot.

Among he Semangs of Malaya and the Battaks of Sumatra, the sun is believed to be the parent of several offspring suns, but

182

is tricked by the moon into devouring them when they threaten to burn up the world.

Since the tool-making hominids emerged some 1.5 million years ago, the foundation stone of "low-energy" architecture was literally laid by the prehistoric cave dwellers. They knew of rock's capacity to store the passive heat of the sun, which was absorbed during the daylight hours, keeping their caves cool, while at night the warmth was preserved by heat emissions from the same source. With the passage of time, increasing trade and cultural interaction between the planter's need to settle down and the inborn migratory urge of the nomad to move on created a neo-cave architecture seen in the numerous cave stations along the so-called silk routes in Central Asia, India, and other locations from the east coast of China to the Mediterranean. As the caravans traversed vast continents, these wayside cave stations, excavated in almost-vertical rock escarpments, provided travelers with much needed rest and recreation as they paused during their long and arduous journeys.

In the vast area of

---

180. Hadrian's Pantheon in Rome (117–126 A.D.) is a "glorified nomadic yurt." The sun is represented by the circular opening in its rotunda dome. Courtesy: National Gallery of Art, Washington, D.C. Painting by Giovanni Paolo Pannini, c. 1740.

181. The sun and the moon peep into the window of a gondolier's home in Venice.

182. The hilltops of mountain escarpments in which caves were excavated served as blueprints of architectural designs of temples and shrines, such as this 11th-century sun temple at Modhera, India.

Central Asia where stone was scarce, the nomadic societies imitated the cave in the shape of a mobile circular tent, called a *yurt*, made of skins and cloth. And just as the prehistoric caves were the forerunners of the magnificent cave temples along the silk routes, the yurt also eventually developed into magnificent edifices, such as Hadrian's Pantheon in Rome and its rotunda (117–126 A.D.). In this "glorified yurt" the large circular opening for light, warmth, and ventilation in its huge dome is obviously a replica of "the sun" in the roof of a nomadic tent. These "containers" of energy — as the architect and historian Lewis Mumford visualized them in a modern context — varied from the Mesopotamian habitats of a "hole dug out of the soil sun-dried to brick hardness, to mud and other types of huts." Other Neolithic containers included stone and pottery utensils, structures such as barns and granaries, and collective containers such as irrigation ditches and villages themselves.

The low-energy architecture of the earliest urban settlements is shown by the ruins of the Indus Valley and Harrapa, in the divided Punjab of India and Pakistan, where solar energy was effectively utilized as early as 4000 B.C. The orientation of the buildings shows the awareness of their citizens that the sun favors the southern and neglects the northern exposures of buildings; the axis and positioning of the house and the space within determines the amount of sun it receives. This method of planning is also evident in the 1st and 2nd century A.D. ruins of Sirkap, near Islamabad in Pakistan, where the north-south direction of the streets, lined on both sides with brick houses, provided every house with adequate sunshine. An altar for worshipping the sun was located at the top of the settlement.

183. Since tool-making hominids emerged some 1.5 million years ago, the foundation stone of "low-energy" architecture was literally laid by prehistoric cave dwellers. Rock's capacity to cool caves by storing the passive heat of the sun during the day and to keep them warm at night, was effectively used in cave temples carved out of solid rock, such as this one at Petra (in modern Jordan), where remains from the Paleolithic and Neolithic periods have been discovered.

The Zoroastrians, who worshipped fire as the representative of the sun on Earth, built their altars either in the shape of a circular design (*kukeldash*) or in a crossed circle (*shashtepa*). These altars were central to their sun-inspired circular urban planning — as seen in many Zoroastrian townships in Central Asia — whereby each house lining the circular streets receives sunshine at one time or another. This Zoroastrian architectural plan was also adapted by the Greeks and the Romans with suitable modifications — for example, the medieval Roman town of Bram in France, founded in 333 A.D.

Traditional cultures commonly believe that exposure to the sun nurtures good health, and so a great deal of emphasis was laid on how sunlight is used for sanitation and health. As in the Indus Valley, the very elaborate building codes, including health care and sun-related sanitation engineering, were also applied in ancient Egypt, dating from the Middle Kingdom (2100–1700 B.C.) where sunshine was effectively used for warm-water bathroom facilities and sewage systems. Roman architecture profited from the ideas of medical authorities, such as Oribasius, who writes: "south-facing areas are healthy places because of their exposure to the sun." He also notes that north-facing areas are the least healthy, because they "do not receive much sun and when they do, the light falls obliquely without much vitality."

As with the arts in general, Greek low-energy architecture benefited a great deal from the ideas assimilated from Egypt through the Aegean in the south, and through widespread contacts with Mesopotamia in the east around the middle of the third millennium B.C. At a time when Syria and the Levant were becoming highly urbanized, there is evidence that the Egyptians and Nubians were trading as far away as eastern Persia, whose low-energy architecture had been influenced by the Indus Valley buildings designed to stay cool in the blazing desert sun. The Egyptian city of Iwnw near Cairo — later called Heliopolis by the Greeks — was already a major center of learning on subjects including solar architecture. It

184

185

184. Prehistoric caves were the forerunners of the cave stations along the arduous silk routes. This 4th-century Maijishan cave temple in China, excavated in an almost-vertical rock escarpment, is one such station.

185. The architecture of the magnificent silk-route monoliths has very little in common with prehistoric caves, except for the technique of using the sun's passive heat.

186   187

was primarily the Nile-dwellers' effective utilization of solar energy that prompted Gior-
dano Bruno, the champion of Renaissance
humanism at the end of the 16th century, to
advocate a return to Egyptian sun worship. His
contemporary, Campanella, who writes about
the "Citta del Sole" (Heliopolis) inhabited by
"solarians with white robes," was another pro-
ponent of heliolithic architecture.

Low-energy architecture became popular
in Greece in the 5th century B.C., when many
parts of the country had been almost totally
denuded of trees, which had been ravaged for
wood to heat homes and for cooking. Wood
was also in demand for smelting operations
and for building homes and ships. Plato com-
pares the hills and mountains of his native Atti-
ca to the bones of a wasted body: "... all the
richer and softer parts have fallen away," he
laments, "and the mere skeleton of the land
remains." By the 4th century B.C., wood had to
be imported huge distances from Asia Minor
and the shores of the Black Sea, forcing many
city-states, including Athens, to regulate the use
of wood and to prohibit the use of olive trees

for making charcoal. Deprived of their portable
charcoal-burning braziers, which kept them
warm in winter, the Greeks living in Olynthus,
Priene, and Delos adapted some of the ideas of
Socrates, as quoted by Xenophon: "... in houses
that look toward the south, the sun penetrates
the portico in winter, while in summer the path
of the sun is right over our heads and above the
roof so that there is shade."

Compared with the sparse vegetation in
Greece, Italy was a lush, heavily forested peninsula in the 3rd century B.C. But so voracious
had become the wood consumption in Rome,
especially as the Roman Empire expanded, that
little was left to fuel industry, to build houses
and ships, and to heat public baths and private
villas. Central heating in the expansive villas of
the wealthy Romans was through hot air produced by burning wood or charcoal in furnaces
called *hypocausts* and then circulated through
hollow bricks in the floors and walls. Since these
could consume as much as 300 pounds (130
kilos) of wood per hour, local wood supplies
were exhausted by the 1st century B.C. and timber had to be imported from as far east as the

*186-187. Just as the design of the rotunda of Hadrian's
Pantheon is based on a nomadic yurt of the kind still
used in Mongolia, passive heat is similarly maximized in
the solar architecture of the Sakuramachi Hospital at
Koganei, Tokyo. By utilizing natural materials such as
wood, earth, and paper, the rooms not only retain the
warmth of the sun but create cozy and stylish living
spaces.*

Caucasus, more than 600 miles (1,000 kilometers) away. The biomass reserves had been so badly ravaged by the 4th century that an entire fleet of ships, the *naviculari lignarii*, or "wood ships," had to be commissioned for the sole purpose of transporting wood from France and North Africa to Ostia, an important Roman port. "Self-sufficiency" became the order of the day, a watchword propagated by Faventinus and later Palladius, the two leading architects of the age.

188

At this critical time in the 1st century B.C., the Romans adopted the low-level heating techniqus of Greek solar architecture. However, unlike the methods used in the small Greek city-states, Roman architects had to take into account the environment of an extended empire, with a whole range of different climates. "We must begin by taking note of the countries and climates in which homes are to be built if our designs for them are to be correct," stated the Roman architect Vitruvius. He explained that "one type of house seems appropriate for Egypt, another for Spain ... one still different for Rome, and so on with lands and countries of varying characteristics. This is because one part of the Earth is directly under the sun's course, another is far away from it, while another lies midway between these two ... It is obvious that designs for homes ought to conform to diversities of climate."

The Roman architects Faventinus and Palladius recommended an ingenious method of hot air circulation that warmed the floors of houses in winter — a technique that had been invented by the Greeks and passed on in the

189

190

188-189-190. *The stump that points the way to the Rainbow Valley farm in the Matakana Hills of New Zealand is a good introduction to the passive solar house built of sandstone with an earth-covered roof blooming with herbs and succulents. The low-energy house uses no toxic materials, and its unique earth-roof is supported by heavy timbers and mud-brick columns. The house's owner, Joe Polaischer, is especially happy with the system by which excess hot water from the wood-burning stoves is directed to under-floor pipes to provide extra heat.*

191

192

193

194

191. The Eco House built by the Wellington City Council in New Zealand is a fine example of the hybrid use of thermal, photovoltaic, and passive solar energy.

192. This townhouse in Bowie, Maryland, is fitted with the newly designed "Standing Seam Roofing" solar electric modules, which emulate conventional standing seam metal roofing. They are made from amorphous silicon PV cells deposited on a thin, flexible, lightweight stainless steel substrate and laminated with advanced polymers onto a flat metal roofing panel.

193. The traditional summer cottages in Scandinavia are now increasingly fitted with solar energy systems, as seen in this house in northern Finland.

194. Because solar energy is the main power source in Kerava, a village near Helsinki, hardly any electricity bills are brought by this postwoman on her daily rounds to deliver mail.

writings of Vitruvius. A shallow pit was dug under a black, heat-absorbent floor and filled with broken earthenware or other rubble. This was overlaid with a mixture of dark sand, ash, and lime. The mass of rubble underneath absorbed large amounts of heat during the day and released it later in the evening when the room temperature cooled. Among the new energy-saving techniques, Palladius also advocated the recycling of bath water and the placement of winter rooms directly above the hot baths so that they would benefit from both the sun's heat and the waste heat rising from the baths below.

The traditional method of using air as the heat-carrying medium is emulated in modern architecture — for example, by OM solar manufacturers in Japan. In its architectural designs, air from the edges of the eaves is drawn up under the roof panels and warmed by heat from the sun's rays. This warm air passes through a duct under the floor space and warms the room. The system, which was launched in Japan by the architect Professor Akio Okumura in 1987, is now being used in some 10,000 facilities from Hokkaido in the north to Kyushu in the south. Another Roman innovation adapted by OM is the attached conservatory or greenhouse in which a glazed extension of the south side of the building is designed to provide a combination of energy gain and additional useful space. Storage may be provided in the thermal mass of the floor and/or walls of the sun-space structure itself, or in the form of a pebble bed using forced and natural circulation.

Tradition dies hard, and in much of the world today dwelling types of ancient or prehistoric origin, so-called vernacular architecture, are still preferred, particularly in the developing countries of Asia, Africa, and Latin America. The effective use of sunshine and warding off its excessive heat remain paramount in the construction of human shelters, and the building materials used are much the same as those used in ancient Rome. They were made of wickerwork plastered with clay on a timber framework, and later of timber and mud brick before more durable materials including stone, marble, and concrete were used in the now famous monuments of antiquity. Even in industrialized countries such as the United States, structures such as barns are built according to old techniques and designs employed in Europe back in the 1st millennium B.C., since experiments

195

196

195-196. The corridor of the environmental education center De Kleine Aarde in Boxtel, Holland, is covered with light-transmitting solar panels that function both for generating electricity and absorbing heat. This kind of double-function system is now used worldwide, especially in the architecture of public utilities buildings with large glass window panes.

and innovations are often more costly than imitation.

Another important passive solar energy innovation around the 1st century was the use of transparent materials to make windows that would let in light but keep out rain, snow, and cold. The Romans had probably come to learn about the use of glass from ancient Egypt and Mesopotamia, where glass beads had been manufactured since about 2500 B.C. The use of glass in Rome was noted in 65 A.D. by the philosopher Seneca: "Certain inventions have come about within our own memory — the use of windowpanes that admit light through a transparent material, for example." Excavations at Herculaneum, Pompeii, and elsewhere pro-vide evidence of glass having been fixed into windows for the homes of wealthy Romans.

As in the past, buildings of today need good external heat insulation and airtight win-dowpanes to reduce heating requirements. Ordinary glass has two important characteris-tics that make it particularly useful in solar ther-mal systems: it allows most visible light to pass through without being absorbed or reflected, but it reflects or absorbs the longer wavelength, infrared radiation, in the form of heat. In low-energy buildings, incoming solar energy is typi-cally stored in a thermal mass, such as a large mass of concrete, brick, rock, water, or a materi-al that changes state (solid to liquid or vice versa), as the temperature changes. The amount

197

197. *A view of the Kushira Information Center in Hokkaido, Japan. In this architectural design by OM solar manufacturers, the traditional method of using air as the heat-carrying medium is employed by passing warm air through a duct under the floor space to warm the rooms.*

198. *The ancient Roman method of under-floor heating is effectively used to store solar energy at the Kushira Information Center by using vertical ducts (yellow) so that the living space remains comfortably warm during winter months, even when the average outside temperature drops to 21° F (-6° C) in this most northerly island of Japan.*

of incoming sunlight is regulated with over-hangs, awnings, and shades, while insulating materials can help reduce heat loss at night or during cold seasons. Vents and dampers are typically used to distribute warm or cool air from the system to the areas where it is needed. In Germany, for example, more than 200,000 square feet (20,000 square meters) of building envelopes are already equipped with transparent insulation.

Today, water heating is the most wide-spread application of solar energy in the world. First developed by a Swiss scientist in the 18th century, most "flat-plate" solar collectors consist of a glass-topped box with a copper absorber plate and small copper tubes through which the heating water flows from bottom to top — usually connected to an indoor storage tank. Solar hot-water systems gained popularity in the United States — especially California and Florida — and Australia at the turn of the

century. In the United States alone, dozens of companies have installed more than 2.5 million solar heating systems while in Japan more than 5 million rooftop solar hot-water heaters have been installed since the mid-1980s. In the early 1990s, environmental concerns in Europe also spurred governments to pay more attention to solar water heating. In the Netherlands some 15,000 solar water heaters have been installed, encouraged by subsidies that are due to be phased out as the industry becomes stronger. The solar hot-water systems and space-heating applications in buildings can be combined in a hybrid integrated system to provide domestic hot water throughout the year while at the same time providing space heating when required.

Israel is an outstanding example of how solar collectors are effectively used on virtually every rooftop in accordance with a law requiring solar water heaters be installed in all new apartment buildings up to eight stories high. As

199. The Akakokko Pavilion — named after the most common bird among some 210 wild species found on the Japanese island of Miyakejima — is a good example of how architectural design can be integrated with nature. The wings of the pavilion, which use solar energy, were constructed without harming the trees, and the large number of people who come to watch the birds do not disturb them in any way.

200. The "Tree House" in Offenburg, Germany, is a bold architectural experiment in the production and use of solar thermal and PV energy. Built in 1994, the 100-kilowatt solar power station generates about 80,000 kilowatt-hours of electricity. The tower of the Hansgrohe solar power station was designed by the architect Rolf Disch to serve also as his residence. It is open on the sunny side and well insulated on the shady side — and water is heated by solar collectors.

201

a result, 65 percent of homes in Israel are currently equipped with solar hot-water heaters, making a total of about 900,000 systems. They provide hot water for 83 percent of Israel's homes, and jobs for thousands of people working in some 30 solar companies. Taking a cue from Israel, most Mediterranean countries now use solar water heaters, and their manufacture has become a cottage industry. In Cyprus, for example, there are more than 150,000 units in operation — one unit for every five individuals, providing the equivalent of 9 percent of the total electricity consumption in the country. A total of 600,000 plants covering an area of 20 million square feet (2 million square meters) have been installed in Greece, where the industry employs some 3,000 workers. In addition to 12 larger companies, there are around 350 "garage manufacturers" that together cover

two thirds of the country's solar energy market. With abundant sunshine at their disposal, the use of solar water heating is fast becoming popular in Asia, Africa, and Latin America, where tens of thousands of systems have been installed, often with little government support.

Solar district-heating projects are also gaining popularity. Sweden leads the world in their development with more than a dozen large projects either built or planned, while in Germany three large projects are under construction. Fixed flat-plate collectors can generate temperatures high enough for most space-heating and hot-water needs, but they are inadequate for many industrial processes, which may require temperatures of more than 570°F (300°C). These higher temperatures can be reached by using tracking devices that change the orientation of the plates so that they face the sun as it moves

201. A new PV-oriented architecture is making its presence felt not only in the developing countries but in the heart of industrialized Europe, as seen in this PV-powered church at Steckborn in Switzerland. It also belies the impression that solar energy is applicable solely in remote and isolated regions of the world..

across the sky during the day, enabling them to capture more energy than fixed collectors.

For electric power generation, equipment such as lenses or mirrors can be used to concentrate sunlight onto either a single collector in a central receiver system or a group of collectors in a distributed system.

Since the mid-1980s, efforts have been under way in a number of countries to encourage people to use solar cookers, as the burning of biomass for cooking in poor countries is one of the principal causes of the destruction of rain forests and the Earth's biosphere. For example, the government of India in cooperation with local self-help groups has promoted a modified box cooker that has been distributed to more than 200,000 families. In parts of China, including Tibet, Gensu, and Qinghai, more than 140,000 parabolic disk solar cookers are now being used, while solar hot-water heaters are also becoming increasingly popular — collectors covering an area of about 10 million square feet (1 million square meters) were installed in 1996. In Pakistan, some 20,000 solar box cookers are being used by Afghan refugees in barren areas such as in Peshawar and Quetta.

In Guatemala's El Progreso region, many families are using box cookers on a regular basis, thus reducing their sizeable expenditure on wood cooking fuel. In the treeless highlands of Bolivia, Freedom from Hunger volunteers have been training local promoters in solar cooker construction and use, and similar efforts have been mounted in Djibouti, Mexico, Zimbabwe, and other countries.

"A place in the sun is one of the most important fundamental rights of every individual," declared the International Congress of Architects, who formulated the Athens Charter in 1940. The renowned architects from all over the world participating in that meeting were inspired by the Justinian Code of Law of 6th-century Rome which stated: "If any object is so placed as to take away the sunshine from a *heliocaminus,* it must be affirmed that this object creates a shadow in a place where sunshine is an absolute necessity. It is in violation of the *heliocaminus'* right to the sun."

202. *A model low-energy house near Aix-les-Bains in France is equipped with a low-temperature solar heated floor, a southeast-facing greenhouse, and a thermal hot water collector.*

203. *An innovative scheme to integrate PV energy with architecture has been introduced by a power company in the Nieuw Sloten district of Amsterdam. The municipality owns the land on which a group of 71 houses has been built, while the power company owns the PV systems installed on the roofs and charges customers only for the use of electricity and not the cost of installation. This cooperative system can supply as much as 200 kilowatts of electricity on a sunny day.*

Nimbus of Photovoltaic Energy

"Give me light so that I may see thy beauty," proclaims the Egyptian Heliopolitan text in homage to the sun. Some 5,000 years later, in 1839, the French physicist A. Edmond Becquerel saw the beauty of light when he discovered that electricity could be produced by sunshine without heat. Until then, the distinction between light and heat of the sun could be explained either in terms of representation, as in the Hellenistic and Roman art in which the sun god Helios and Roman emperors often appear with a crown of rays, or the nimbus of spiritual light emanating from the holy person's head — "the light shining in darkness" as St. John describes the revelation.

In 1876, W.G. Adams and R.G. Day observed the photovoltaic effect in solids — that is to say, the conversion of the energy in light into electricity. Then in 1883 Charles Fritts, an American inventor, made the first solar cells of selenium (an element derived from copper ore) on a brass plate 1 foot (30 centimeters) square covered with a transparent gold film. These early selenium cells converted only about 0.1 to 0.2 percent of the energy in the incident light into electricity, and were too expensive to be used as a power source. It was in early 1954 that a team of scientists at the Bell Laboratories in New Jersey discovered that a silicon device they were testing produced electricity when exposed to sunlight. They went on to make a

silicon cell that converted five or six times more incoming sunlight into electricity than the best previous selenium cell. By the late 1950s, American researchers had developed experimental solar cells with up to 14 percent conversion efficiency. Silicon is the second most abundant element, constituting 28 percent of the Earth's crust, and this was taken as a good omen for silicon solar cells to become a major source of electricity.

Developments in photovoltaic systems accelerated with the need for electrical energy in space vehicles, and the space race between the United States and the Soviet Union led to the first major application of solar cells as a power source in space. These systems were selected for space use because they are lightweight and because they could produce power for long periods of time. The first application of photovoltaic devices as a significant source of power started in 1958, when the second U.S. space satellite, *Vanguard 1*, carried a small array of crystalline silicon photovoltaic cells to power a radio transmitter. These cells worked very well and were thereafter widely adopted in the U.S. space program. Since then, most unmanned satellites and manned spacecraft have relied on photovoltaic cells for all or part of their energy supply. The U.S. space program spawned a number of improvements in cell efficiency and manufacturing techniques resulting in a decrease in prices, and commercial pro-

4 - 205

204-205. Meditating on the tremendous cosmic energy — represented by the nimbus of the one thousand hands of the Buddhist solar deity Avalokiteshvara— a lama is on his way to the 15th-century Himalayan monastery of Thikse in Ladakh using his yak to carry PV panels for the newly installed solar system in the shrine. As in the

Indian part of Himalayas, beasts of burden such as yaks are often used to transport wind and PV systems in some of the remote regions of Tibet in China.

206. A 17th-century Indian sun symbol embroidered with a thread of gold, the "sun metal."

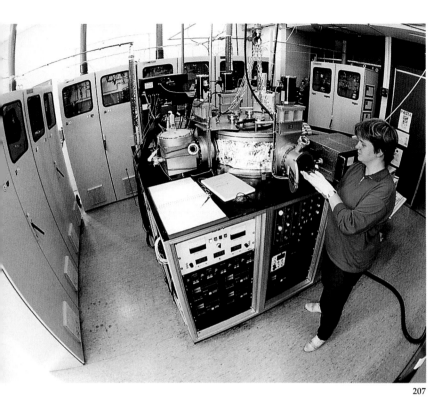

207

208

ducers were able to sell PV systems for a growing number of land-based applications.

While solar cell manufacturing is a complex process requiring advanced technology, many developing countries are now setting up institutions for research in the field of photovoltaics in order to acquire basic electrical and engineering skills. Countries such as India, Brazil, and China, which developed a government-subsidized PV industry from scratch, are fabricating wafers and cells as well as modules. China has seven production lines with a total capacity of 4.5 megawatts. A number of countries are encouraging domestic companies to form joint manufacturing ventures with some of the leading photovoltaic companies in Europe and the United States. Mexico, Morocco, Sri Lanka, and Zimbabwe already have domestic industries putting together photovoltaic panels. In Kenya, Tanzania, and Zambia, the indigenous PV industry still concentrates on installing imported systems, but with outside support they too may soon develop indigenous assembling capabilities.

However, solar electricity is still much too expensive to compete with power from the con-

ventional grid. In order to facilitate widespread use of photovoltaics in places where the electric grid is already in place, the aim of on-going research is to reduce the cost of manufacturing solar cells while increasing their conversion efficiency. The new photovoltaic technology that has captured the most scientific attention and the largest share of the "unconventional" PV market is the so-called thin-film solar cell. A new world record for the most efficient "ultra thin" or "thin film" crystalline silicon solar cell — only 45 microns thick with 21.5 percent efficiency — was set recently by the Australian University of New South Wales at its Centre for Photovoltaic Devices and Systems in Sydney. Produced on a thick substrate, thin films require much less raw material and can be applied directly onto glass or sheets of metal foil, thereby simplifying the manufacturing process and producing a panel that can be easily stored and transported.

A thin-film material that has attracted attention in recent years is amorphous (or non-crystalline) silicon. With a composition similar to common sand, amorphous silicon is inexpensive and can be processed at a low temperature.

207. *A number of deposition methods are being studied for the manufacture of amorphous silicon thin-film solar cells. This device has four chambers for controlling deposition of different types of semiconductor film from which the solar cell is made. This device is also used to develop tandem cells consisting of two or more cells stacked on top of each other to increase conversion efficiency.*

208. *Automatic computerized methods of forming cell matrices have now taken the place of manual methods used until 1996. In this device, cells are placed in a holder in order to form a matrix of 72 cells.*

Originally fabricated by the American researchers David Carlson and Christopher Wronski in the early 1970s, its first commercial application in consumer electronics devices was introduced by the Japanese in the following decade. The system has been considerably improved by Stanford Ovshinsky at his Michigan-based company Energy Conversion Devices, Inc. (ECD) in cooperation with the Japanese company Canon. This system combines three layers of amorphous silicon by "doping" each layer with a chemical that allows it to absorb specific wavelengths of sunlight, thereby capturing more of the sun's rays while minimizing the efficiency decay to roughly 10 percent. With Canon's backing, ECD plans to fabricate panels using large sheets of stainless steel as a substrate topped with a protective layer of Tefzel, a type of Teflon. Unlike glass, which is heavy and breakable, steel shingles are lightweight and flexible.

Research in amorphous silicon solar cell technology has accelerated in Europe as well. At the University of Utrecht in the Netherlands, the cell has been tested for application on flexible films. Elsewhere in the country,

researchers at the Technical University of Delft have been conducting a fundamental investigation into organic solar cells involving titanium dioxide pigment. This is essentially based on the cell invented at the Swiss Federal Institute of Technology that is made of a dye containing ruthenium and titanium oxide, and that functions in much the same way as photosynthesis in plants. Unlike solar cells, which convert sunlight into electricity, in this electrochemical cell, photosynthetic organisms use solar energy to produce carbohydrates from water and carbon dioxide. Scientists believe that the cost of this "nanocrystalline" PV cell will be one fifth that of today's cells, an estimate that has led two large corporations, Asea Brown Boveri and Sandoz to collaborate with the Swiss institute.

One approach being pursued by researchers is the development of highly efficient solar cells — made from materials such as gallium arsenide — that have already achieved efficiencies of more than 30 percent in the laboratory. Such cells could be used in conjunction with lenses and reflective mirrors that focus the sun's rays onto them, greatly reducing the

209. Amorphous silicon solar cell technology has been tested for application on flexible films at Energy Conversion Devices, Inc., in Troy, Michigan. The flexible carrier can replace the heavy glass substrates on which the solar cell structure was originally developed.

210. The picture shows a working flexible cell prototype that was made at the University of Utrecht, Holland. It will be possible in the future to manufacture flexible solar cells more cheaply and incorporate them more easily into applications in roofing and outer wall elements and other consumer products.

amount of semiconductor material needed. The array is usually mounted on a special dual-axis tracking device that allows it to be continuously pointed toward the sun. Concentrator PV cells achieve somewhat higher conversion efficiencies than nonconcentrating cells. In 1991, Boeing achieved a conversion efficiency of 32.6 per-

cent — the highest recorded for any type of PV device — with a multijunction cell under light concentrated 100 times. Because concentrators are unable to use diffuse sunlight, they are most effective in areas of high and direct sunlight, such as deserts and other arid regions.

211. All the houses in the solar township of Kalil in Israel are equipped with individual solar systems composed of PV panels and thermal collectors for water heating. As seen in this hybrid system installed on the rooftop of a house, the solar water collector is placed beneath the PV panel to absorb the heat from the PV system that would otherwise reduce its efficiency.

212. As in space, photovoltaic systems for generating electricity are indispensable in remote locations such as Antarctica. The New Zealand Antarctic Programme also uses solar energy to power batteries for field radios and other scientific research equipment.

Photovoltaic Lifestyle

The 1973 Arab oil embargo gave new impetus to the photovoltaic industry, as government agencies dealing in energy and scores of private companies invested billions of dollars in advancing the state of photovoltaic technology. By 1980, the efficiency of commercial PV modules had risen to more than 10 percent and their price had fallen considerably. Since the 1980s, solar cells have been widely used in telephone relay stations, microwave transmitters, remote lighthouses, and roadside call-box applications, even when conventional power sources are available. In Europe, thousands of light buoys and beacons have been installed in which the batteries for lamps are charged with PV systems. Photovoltaic-powered weirs are being installed alongside integral water management, automation, and measurement facilities. Photovoltaic-powered lamps are being used by municipalities at public works, bus stops and shelters, information billboards, telephone booths, road-crossing warning signals, parking meters, and a host of other cost-effective applications — not only in remote places but in central urban areas.

More than half the solar cells manufactured worldwide are now used in consumer products requiring between a few milliwatts and a few watts of power. By the late 1980s, Japanese solar cells were being used in millions of calculators, watches, and battery chargers each year. Since then, the Japanese have sold an average of about 150 million such devices annually, an application that accounts for 4 megawatts of solar cells, or 7 percent of the global market. Japan's environmental agency recently reached an agreement to replace 10 percent of the nation's soft drink vending machines with models powered by solar batteries. The accord reached between the agency and domestic vending machine makers called for the introduction of about 200,000 new machines within seven years, officials at the agency said. According to the agency, Japan had more than 4.5 million vending machines at the end of 1996, 2 million of them selling soft drinks.

In rural areas, too, new photovoltaic devices for threshing, drying, and storing grain in silos are being employed in agriculture. Thousands of PV-powered drinking troughs for cattle are now in use across Europe — the system powers a pump that conveys water from an irrigation ditch into a trough fitted with a battery and control system that ensures that the trough remains full, even during overcast weather. PV-powered water pumps have become very popular, especially in the developing countries, where tens of thousands of them can be seen in the remotest of regions. New ideas are emerging to increase agricultural production and preservation; on the island of Miyako, Japanese farmers are using numerous PV-powered insect killers in their fields instead

---

213. *Remote places that cannot be connected to the grid can now be supplied independently with electricity by new solar modules such as this one designed by Siemens for standard television sets, radios, and refrigerators.*

214. *Photovoltaic artwork by Gliesdorf Styria in a school yard (2.2-kilowatt modules).*

215

216

217

218

219

220

22

215. Like the millions of miniature PV systems used in small calculators, watches, and other electronic devices, medium-size PV equipment is now being increasingly used in transport facilities, such as this information board at a remote location in Europe.

216-220. A variety of solar lamps and mobile PV lanterns in different sizes and designs are being manufactured and used in a number of countries.

221. A woman sells her wares beneath a PV streetlight in a hamlet of Mande Province in Mali.

of harmful insecticides. Farmers and fishermen in India are using solar lanterns connected to one or two fluorescent lamps for night work, the batteries of which can be charged with a small PV panel. Electrical appliances for which farmers previously used diesel generators can now be run from a photovoltaic power generation system for battery charging — a 40-kilowatt prototype station started operating in Thailand in 1995 and is now being used by about 500 households. It can charge 75 car and other kinds of batteries in one day.

Photovoltaic-powered vehicles are still a novelty, although the cars participating in the World Solar Challenge have progressively improved their performance since 1987, when the triennial 1,860-mile (3,000-kilometer) race — starting from Darwin in the north of Australia to

222

222. *A number of entrepreneurs in India are now manufacturing solar lanterns, the batteries of which are charged with PV power. Here, two agents of an electronics company in the southern state of Kerala are about to take off in a boat to deliver two solar lanterns to offshore fishermen, who usually fish at night.*

Adelaide on the south coast — was inaugurated. In 1996, Honda's "New Dream" attained an average speed of 55.77 miles per hour (89.76 kilometers per hour) and succeeded in beating the record of 41.58 miles per hour (66.92 kilometers per hour) set by General Motors' "Sunraycer" in the first race.

The increasing public concern for the protection of the environment has led several automobile manufacturers and engineering schools to build electric-powered vehicles. The National Institute for Environmental Studies in Japan recently displayed the electric-powered "Luciole" that was showcased along with Honda, Mazda, and Nissan models at the Tokyo show. General Motors has produced the EV-1, whose batteries can be charged with solar energy. Other car manufacturers have come up

223

223. Thousands of light buoys and beacons in the United States, Europe, and Japan have been fitted with PV systems that charge the batteries with which the lamps are powered.

224   225

with low-pollution cars with hybrid engines running on both gas and electricity, such as Toyota's RAV4 EV and Peugeot's 106. Photovoltaic systems are especially useful in a variety of electrically propelled boats, as there are fewer space constraints. The 120-day voyage of Kenichi Horie — who covered a distance of 10,000 miles (16,000 kilometers) from Ecuador to Japan in a PV-powered boat made from 22,000 recycled cans — is not only a feather in the cap of the Japanese adventurer, but convincingly demonstrates the far-reaching power of his solar-propelled *Malt's Mermaid* — in much the same way as Thor Heyerdahl's *Kon-Tiki* voyage belied the skeptics about the power of wind energy.

Aesthetic concerns were aroused by the unseemly and haphazard manner in which more than 200,000 PV panels have been mounted on the rooftops of homes in Mexico, Indonesia, South Africa, Sri Lanka, and other developing countries over the past decade. More attention is now paid to photovoltaic architecture by both governments and private agencies to integrate and aesthetically harmonize the designs of houses and commercial buildings. This has launched a new trend in architectural design, the result of which can be seen in thousands of households worldwide. For example, the nonprofit group Asociación para el Desarrollo de Energia Solar has installed more than 2,000 rooftop PV panels in the Dominican Republic during the past nine years. Since 1979, similar organizations have been established in China, Honduras, Indonesia, Sri Lanka, Zimbabwe, and elsewhere.

Worldwide research is also continuing on making photovoltaic products that enhance the beauty of architectural designs and protect the environment. The impetus has partly come from the Global Environment Facility (GEF), a fund set up under the joint management of the World Bank, the United Nations Development Programme, and the United Nations Environment Programme to finance projects that, although not strictly economically viable today, will in the long run benefit the global environment by reducing carbon dioxide and other pollutants in the atmosphere. In Zimbabwe, a 7-million-dollar GEF grant approved in 1992 is financing a revolving fund for electrifying 20,000 households in five years.

224-225. *The far-reaching power of solar energy and its environmental benefits were demonstrated by the Japanese Kenichi Horie, who, in 1996, covered a distance of 10,000 miles (16,000 kilometers) from Ecuador to Japan in a PV-powered boat called Malt's Mermaid, made from 22,000 recycled cans.*

226

227

228

229

226. PV systems are now increasingly used in automatically operated locks, especially in the Mediterranean and Benelux countries. The systems also provide energy for the signals and for illuminating stations at night.

227. Electrically propelled boats using batteries charged with PV panels have become popular not only because they free users from dependence on charging stations, but also because they are quiet and can enter sanctuaries where gasoline-fueled motorboats are prohibited.

228. Ferries are now a common sight on busy waterways in many countries in Europe and the United States, like this ferry used by cyclists between the villages of Suwâld and Garyp in Holland.

229. An auxiliary photovoltaic power system on a motorboat off the coast of Boston in the United States.

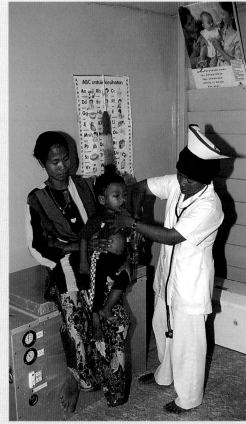

230

231

232

233

234

230-238. *Piamenggil, the isolated island of fishermen off the east coast of Malaysia in the South China Sea, received a new lease on life when a 10-kilowatt PV system was installed by Siemens. Since then, its 300 or so inhabitants have come in contact with the outside world through a system of solar telecommunications. Almost all the homes have solar lights, some have fans,* *and there is a television set in the community center. Electricity is also benefiting children at school and a clinic to which people can come day or night. Some enterprising fishermen have even opened up rest houses and a food kiosk for tourists who have started coming to the island in large numbers.*
The pictures show: (230) Housewives watching television

235

236

237

238

at the community center. (231) Women preparing dinner by the light of a solar lamp. (232) A nurse examining a child in the clinic while his mother sits on a solar refrigerator. (233) Tourists arriving from Singapore. (234) Fishermen mending their nets beneath a solar streetlight before sailing away for night fishing. (235) A view of the island and the PV antenna for telecommunications. (236) The Piamenggil pier with PV lampposts. (237) Children playing with coconuts under a PV streetlight. (238) Fishermen eating hamburgers at the recently opened kiosk.

239

In India, a 42-million-dollar World Bank loan and GEF grant will support a program of 2.5-megawatt aggregate capacity photovoltaic systems including solar lanterns. Similar World Bank loans and grants are available for China, Indonesia, Sri Lanka, the Philippines, and other developing countries in order to strengthen nascent photovoltaic industries that could help in creating economic opportunities and jobs in rural areas. The grant to Zimbabwe, for example, is supporting six small photovoltaic supply companies and a larger enterprise that import cells and assemble them into commercial panels.

The University of New South Wales' Centre for Photovoltaic Devices and Systems has developed a solar roof tile with a "static concentrator" that incorporates solar cells that respond to light falling on both their front and rear surfaces. The design uses a three-dimensional concentrator, which allows the fitting of

the module within normal roof tile constraints. The aim is to make aesthetically pleasing and cost-effective photovoltaics an integral part of homes and commercial buildings. A company in California is working with utilities to market more traditional-looking solar shingles, blue roof modules, and prototype solar panels incorporated directly over the rubber or plastic roofing membranes of flat-roofed buildings. In addition to developing the next generation of solar roofing, some companies are incorporating solar cells into exterior cladding for commercial buildings. Called "Power Wall," the cladding is made in panels that are up to 4 by 5 feet (1.2 by 1.5 meters) in size and that cost approximately the same as other premium cladding materials, such as granite.

The façade of a recently completed grid-connected photovoltaic and thermal system of the Pompeu Fabra Library at Mataró in

240

241

239. *A number of south-facing sound barriers in Europe are now lined with grid-connected PV panels, such as this 100-kilowatt solar system installed on a motorway near Chur, Switzerland, as a solar-powered refrigeration truck passes by.*

240. *PV systems are well suited to providing cost-effective power for public facilities that require small amounts of energy, such as this bus stop integrated with PV cells.*

241. *This movable warning sign is another example of the multiple applications of PV systems, even where grid-connected electricity is available.*

242

Barcelona is made entirely of multifunctional PV-thermal modules containing blue polycrystalline silicon solar cells that are semitransparent. This rectangular building faces south, and a semitransparency of 15 percent is obtained with horizontal transparent slits of about 1 inch (2 centimeters) each that extend longitudinally throughout the 2,400 square feet (225 square meters) of the photovoltaic façade. The four rows of tilted skylights allow diffused light to enter on the north side, and some of the roof modules are also made of semitransparent amorphous silicon. This multifunctional architectural concept in which the library will supply its surplus energy to the grid was first proposed in 1991 by Dr. Antoni Lloret to the European Community's THERMIE Programme, and has been realized by the Catalan company TFM with the support of ZSW of Stuttgart and the University of Barcelona.

Researchers at Johns Hopkins University have also been working with similar films that not only appear to be cheaper to produce than crystalline cells, but can also be transparent — making them appropriate for a whole class of building-integrated solar products such as windows.

A very popular practice in several countries today is to install relatively small solar arrays on the roofs of residential and commercial buildings in areas already connected to a power grid; these can be used to meet either all or part of a building's power needs. At times when a solar array produces more power than its users need, the excess electricity can be fed into the grid or stored in batteries on site; and when the array produces little or no electricity, the users can either draw power from the grid or from their storage system. This arrangement is beneficial for homeowners, who are paid for their surplus power, and the utilities profit by

242. *In anticipation of the widespread use of electric vehicles, parking shelters with PV rooftops, such as this one in California, are being constructed so that drivers can park and charge the batteries of their electric vehicles (EVs) while they go out shopping.*

saving a great deal of money they would otherwise spend on new power plants to meet peak demand. Moreover, unlike central power stations, which require large tracts of land, rooftop PV systems need no land other than the area already occupied by the buildings. Although the generating capacity of a typical rooftop system is probably relatively small — up to a few kilowatts — very large amounts of electricity could be generated if rooftop systems were widely installed in cities in sunny climates.

Community photovoltaic installations in remote areas and on small islands have been very effective. A particular case in point is the fishing community of Piamenggil, off the coast of Malaysia, where an entirely new lifestyle has emerged since a 10-kilowatt photovoltaic system was installed by the German electronics company Siemens about five years ago. Since then, this community of some 300 inhabitants living in complete isolation in the South China Sea has come in contact with the outside world through radio and television. They now have two or three lights in each of their homes, and electricity is also benefiting their children at school and in a clinic. The project was pioneered by a private enterprise with only limited support from the government and aid agencies.

In the industrialized countries, too, remote regions without grid electricity connections are benefiting from photovoltaic systems, such as the La Garrotxa district in Catalonia, Spain, where a large number of dwellings scattered among the valleys and mountains lack most of the basic modern services, such as electricity and telephones. In this region the local authorities promoted a very effective electrification project within the framework of the European Community's THERMIE Program, in which 65 sites were chosen for installing stand-alone PV systems. This was done in cooperation with an association of users that is now responsible for

243

243. The large-scale use of PV systems is inevitable in public facilities like this bus shelter lit up at night with PV power, which also illuminates the time-schedule board of bus services (right) and changes the timing automatically.

244

245

246

the operation, monitoring, and maintenance of the plants. Users are trained in handling the appliances and load management, and a monthly fee paid by the association's members partly meets the cost of maintenance tasks.

Even in Europe's grid-connected highly populated areas, tax incentives, low-interest financing, and cash rebates have tempted building owners to install photovoltaic systems. The Thousand Roofs Program launched in Germany in 1990 has since been extended to include 2,500 roofs. Switzerland aims to place at least one PV system in each of the country's 3,029 villages by the end of this

244. There is no way that grid-connected electricity can reach the vast isolated regions of Australia, where the only alternative is to install PV-equipped telephone booths such as this one.

245. Portable PV systems for work and recreation are becoming popular worldwide — like the solar modules this tourist is using on the isolated and "bewitched islands" of Galapagos.

246. The light in this PV-powered insect trap is switched on automatically after sunset and goes off three-and-a-half hours later. It attracts insects, which collide with the metal grid and drop into the receiver below. Hundreds of such insect traps have been installed on the Japanese island of Miyako, where farmers prefer using them rather than harmful insecticides.

247

248

decade. The Netherlands is planning to install 250 megawatts of photovoltaic systems by 2010. Norway already has 50,000 PV-powered country homes, and an additional 8,000 are being "solarized" each year. Austria, Switzerland, Spain, and Denmark are preparing their own incentive programs. Japan, meanwhile, has devised a strategy to install some 62,000 building-integrated PV systems with a combined capacity of 185 megawatts by the end of the 1990s. With their peak power demand coming from air-conditioning use on sunny days, Japanese companies are concentrating particularly on developing solar-assisted air-conditioning units. A government subsidy that initially covered half the cost of such systems was reduced to 30 percent in 1997 and will decline gradually until it is eventually eliminated.

249

That photovoltaic solar energy could now compete with energy generated from oil, gas, and coal is evident from the Enron Corporation's plans to build a 150-million-dollar plant in the southern Nevada desert that would be the largest in the United States. Enron's 100-megawatt plant would be more than a dozen times the size of any other that employs photovoltaic or solar power cells. The company already has preliminary support from the U.S. Department of Energy, which tentatively plans to buy Enron's solar power output as long as the rate is competitive with power from conventional sources. The plant is expected to begin operating by the year 2000, producing enough electricity for a city of 100,000 people. During the last decade, the cost of solar power generation has fallen by two thirds. The Worldwatch Institute, an environmental group in Washington, D.C., stated recently that solar cell electricity, which now costs as little as twenty cents a kilowatt-hour, might fall to ten cents by 2000 and four cents by 2020.

247-249. *La Garrotxa district in Catalonia, Spain, is among the 65 sites chosen by the European THERMIE Programme for the installation of stand-alone PV systems. This is done in cooperation with an association of users, who become responsible for the operation, monitoring, and maintenance of the plants. The pictures show: (247) A view of the Garrotxa Valley. (248) A farmer's house with a PV panel on its roof. (249) Children at lunch at the Garrotxa community center.*

250-251 *The Pompeu Fabra Library at Mataró in Barcelona reflects a multifunctional architectural concept in which semitransparent PV modules are integrated to produce peak thermal and electrical power of 53 kilowatts, which is grid-connected. The front of the building is made of polycrystalline cells, and the PV panels on the roof are made up of cells encapsulated in opaque and semitransparent modules. The four rows of tilted skylights allow diffused daylight to enter and illuminate the building's interior.*

250

251

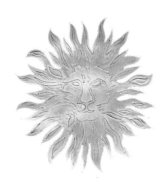

Cosmic Energy of Hydrogen

The lightest of the elements, as well as the most abundant, hydrogen is a hydrocarbon without the carbon, since when combined with oxygen to produce heat or electricity, its main by-product is water. The gas could be produced by water electrolysis from clean solar power — by electricity obtained from photovoltaic panels or generated by hydropower, wind turbines, or with the help of other solar energies, such as tides and ocean waves. The basic idea is to produce sufficient solar electricity during summertime and use it for electrolytically decomposing water into hydrogen in a pressurized electrolyzer, and then to store the hydrogen in a pressurized vessel. The hydrogen thus stored can be converted back into electricity during the wintertime through a fuel cell, while the water produced in the fuel cell is circulated back to the electrolyzer, thus closing the circuit. Hence hydrogen in its pure form has a much better prospect of becoming the ultimate carbon-free energy source of the future.

The prospects are bright for hydrogen being used as car fuel and thus becoming the environment's savior from the menace of suffocating pollution. This was demonstrated by a prototype car, "NECAR II," which was tested recently in Berlin by Daimler-Benz. The idea of using hydrogen gas as fuel dates back to 1807, when one Isaac de Rivas in Paris patented a hydrogen gas–powered vehicle. The valves and

ignition of its engine were operated by hand, but problems with the timing mechanism proved difficult to overcome. "NECAR II" has its own problems, since fuel cells that produce hydrogen are bulky and expensive, and there is much work to be done before the car goes into production. But its feasibility has been proven already as the car is fueled by the chemical combination of hydrogen and oxygen produced by 300 fuel-cell plates that make electricity on board — the only emission from the exhaust is water vapor. With an electric motor in front, it has a two-speed automatic transmission and can be driven at a speed of 62 miles per hour (100 kilometers per hour) — and if all goes well, hydrogen-powered cars may well be on the road by about 2005.

Another fuel-cell application has been announced recently by the U.S. Department of Energy and Arthur D. Little, Inc. They have produced electricity from gasoline that yields twice as much useful energy per gallon as a present car's engine does and with 90 percent less pollution. The fuel cell that will make these cars practical and economical enough to manufacture and market could also be effectively used in the production of solar energy on a much wider scale. A fuel cell is an electrochemical producer of electricity in which continuous operation is achieved by feeding fuel and an oxidizer to the cell and removing the reaction products. Sir William Grove constructed such a cell in 1839,

252. *Parked in front of the quadriga of the sun god Helios on the Brandenburg Gate, this prototype hydrogen car was tested recently in Berlin by Daimler-Benz. "NECAR II" is fueled by the chemical combination of hydrogen and oxygen produced by 300 fuel-cell plates, which generate electricity on board. The only emission from the*

exhaust is water vapor. With an electric motor in front, the car has a two-speed automatic transmission and can be driven at a speed of 62 miles (100 kilometers) per hour.

253. *The lion face of this French sun symbol represents solar power.*

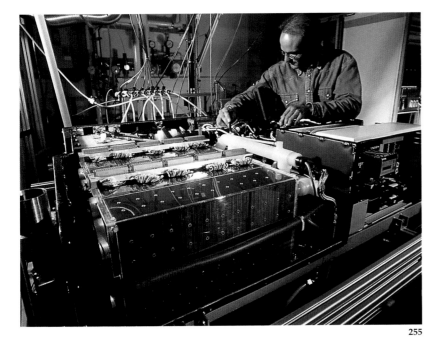

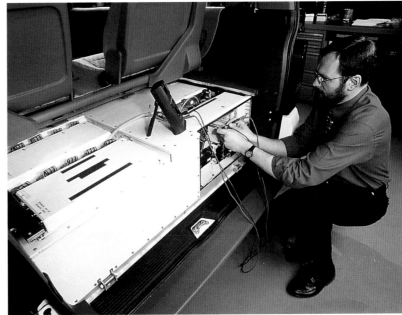

with hydrogen and oxygen reacting on platinum electrodes to produce electricity. One of the main objectives of electrochemical research had been to find a process for the direct electrochemical oxidation of fuels such as coal to supply the growing requirements for electricity. However, the interest in fuel cells dwindled when large generators driven by steam power plants and water power were developed.

But then the need for silent portable electric generators to power electronic equipment during World War II arose, and this created renewed interest in fuel-cell research, which was subsequently boosted when efficient energy-conversion devices were needed for the space program. This resulted in the development of the hydrogen-oxygen fuel cell for manned space travel and its use in converting hydrogen into electricity through solar energy at an efficiency as high as 65 percent.

The romantic post–World-War-II view in the early 1950s that nuclear fission energy would be the answer to all our energy requirements rapidly faded, not only because of the uncompetitive cost of nuclear power, but primarily due to concern about accidents and the storage of nuclear waste. So now some scientists are pinning high hopes on the development of "environmentally safe" nuclear fusion (instead of fis-

sion) as a competitively priced energy source.

The basic fuel used in government-funded fusion research programs in the United States and Europe — the Tokamac project at Princeton University and the Joint European Torus (JET) — is deuterium, a form of hydrogen that is easily extracted from water. Energy is produced when light atoms of hydrogen are fused together to form heavier ones, such as helium. Happily, deuterium produces none of the greenhouse gases or acid rain emissions associated with other forms of energy production. However, it is presumed that a fully operational nuclear fusion power station could be producing power only until the middle of the next century, when world energy demand is expected to have doubled and when fossil fuel reserves are likely to have been exhausted. Another snag is that even in these comparatively "safe" nuclear fusion plants, deuterium has to be mixed with tritium, a radioactive form of hydrogen. After spending billions of dollars, scientists may well end up with a hybrid system with radioactive nuclear health hazards.

But hydrogen in its pure form is another matter. It is not a primary energy source like oil or coal, but rather a "clean" energy carrier like electricity, which can be stored and converted into different forms of energy with the help of heliotechnology. Until now, hydrogen has been

254. At Princeton University, researchers are refitting a doughnut- shaped machine called Tokamak. Using powerful magnets to confine a superhot ring of gas, Tokamak heats the nuclei of hydrogen isotopes extracted from water until they fuse together, releasing energy — but not without radioactive nuclear health hazards.

255-256. Technicians working on the "NECAR II" engine are trying to improve the fuel cell to make the car practical and economical enough to manufacture and market. A fuel cell is an electrochemical producer of electricity in which continuous operation is achieved by feeding fuel and an oxidizer to the cell and removing the reaction products.

used solely for making high-tech rocket fuels and for chemical reactions in hydrogenation. Its use in conventional technology is for oil refining, synthetic fuels, in the electronics industry, in the food hydrogenation industry, weather balloons, and so on. These niche markets and applications are the essential pathway towards the wider utilization of hydrogen technology.

However, hydrogen is not used today for direct energy applications, although the logic of a transition to hydrogen technology has been put forward by scientists for more than a century — in the 1870s, Jules Verne wrote about hydrogen being a good substitute for coal.

Currently, large quantities of petrochemicals are being used in everything from fertilizers

257

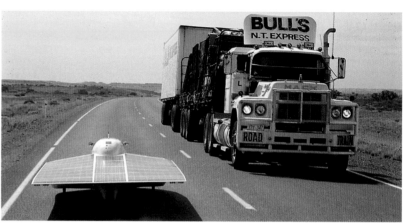

258

259

260

257. *Until the hydrogen car can be produced commercially, the only alternative to protect the environment is to use electric cars whose batteries can be charged with solar energy derived from PV or wind power supplied by wind turbines, such as these at Punta de Tarifa in Spain.*

258. *Photovoltaic-powered vehicles are still a novelty, although the cars participating in the World Solar Challenge have progressively improved their performance since 1987, when the triennial 1,800-mile (3,000-kilometer)*

race — *starting from Darwin in the north of Australia to Adelaide on the south coast — was inaugurated.*

259. *Electric cars and scooters have already appeared on French roads, such as these in Strasbourg.*

260. *1n 1996, Honda's "New Dream" attained an average speed of 55.77 miles per hour (89.76 kilometers per hour) and succeeded in beating the record of 41.58 miles per hour (66.92 kmh) set by General Motors' "Sunraycer," which competed in the first race.*

to cleaning fluids and clothing. These carbon-based materials are derived from the various oil by-products that flow in enormous quantity from the world's refineries. If oil consumption were abandoned, natural gas could be converted into hydrogen and carbon, with the latter going into useful materials rather than into the atmosphere. In the long run, natural gas could produce petrochemicals from crops on a sustainable basis, which would be a far more important use of cropland than simply having it produce fuels. U.S. scientists have also proposed producing hydrogen from switchgrass grown on marginal cropland, with plans to use it in fuel cells that provide farms with electricity and heat. Technological developments have

261

262

263

264

261. *The increasing public concern for protection of the environment has led several automobile manufacturers to build electric-powered vehicles. General Motors has introduced the EV-1, while other car manufacturers have produced low-pollution cars with hybrid engines running on both gas and electricity, such as Toyota's RAV4 EV and Peugeot's 106.*

262. *The "Solectria Sunrise" electric vehicle, powered by Ovonic nickel-metal hydride batteries, set a new*

distance record of 375 miles (603 kilometers) on a single charge in the 1996 American Tour de Sol race.

263. *As part of their interscholastic program, Hawaii Ka'ahele La' students at Roosevelt High School in Honolulu are working on a PV-powered electric vehicle.*

264. *Environmental concerns prohibit entry of any vehicles onto most golf courses except electric cars such as these.*

made it possible to convert biomass into gaseous hydrogen at a higher overall conversion efficiency than is feasible by converting it to liquid fuels such as methanol or ethanol. In this way, biomass-derived hydrogen would complement hydrogen produced from solar and wind energy.

Hydrogen is an attractive option because it can be stored and transported over long distances. Hence, as natural gas supplies level off, the fuel most likely to replace it is hydrogen, the simplest of the chemical fuels. It could be carried through natural gas pipelines already in existence in many parts of the world and then converted into electricity through fuel cells. This would be more efficient than the oil or electricity distribution systems in place today. Power lines, for example, are expensive to build, and over a distance of 620 miles (1,000 kilometers) lose between 3 and 5 percent of the electricity fed into them. In the early stages of the transition to hydrogen, the new energy gas could be added to natural gas pipelines in concentrations

of up to 15 percent to form a mixture known as hythane. In the long run, it would not be too difficult to modify today's natural gas pipelines so that they could transport hydrogen. Hydrogen could also be produced from natural gas, either in central facilities or right at the gas station.

In Israel, research is being done at the Weizmann Institute into the methods of utilizing solar energy in chemical form. The most developed concept so far is the "chemical heat pipe," based on the reversal reaction through which methane is converted into hydrogen and carbon monoxide. The reaction is conducted in a solar furnace by passing methane and steam or carbon dioxide through a catalyst bed at 1,400° to 1,800°F (800° to 1,000°C). The gaseous reaction products are cooled, stored at high pressure, and transported to the point of use. The stored solar energy is released in a "methanator" plant that uses the reaction between hydrogen and carbon monoxide to produce methane, which at the same time generates high temperatures that are used to produce steam, electric

265

265. Research is being done at the Weizmann Institute in Israel into methods of converting methane into hydrogen. A different approach is being taken in Japan at Osaka University, where tests are being conducted to produce hydrogen using green algae and photosynthetic microbacteria.

power, or both. The methane is returned to the solar site, completing the cycle.

Another method of transporting stored hydrogen would be to chemically convert it into a liquid and then convert that into electricity at its destination. Additional hydrogen could be produced in individual homes and commercial buildings using rooftop solar cells, and it could either be stored in a basement tank for later use or be piped into a local hydrogen distribution system. In either case, it would gradually fill the roles occupied by oil and natural gas today, including home and water heating, cooking, industrial heat, and transportation. The gas could also be produced at large wind farms and solar ranches established in sunny and windy parts of the world, since they could generate electricity to be fed into the grid when power demand is high and could produce hydrogen when it is not.

Indeed, the fuel cell may one day be thought of as the silicon chip of the hydrogen economy. It could become the standard equipment of industry, and homes could use reversible fuel cells capable of producing hydrogen from electricity and vice versa. One such example is the energy-self-sufficient solar house (SSSH) in Freiburg, Germany, built by the Fraunhofer Institute for Solar Energy Systems in 1992. It has no grid connection, and its entire energy requirement for heating, domestic hot water, electricity for lights and all other appliances including cooking is supplied by solar energy. Surplus energy from the PV generator in summer is fed into an electrolyzer to decompose water, and the resultant oxygen and hydrogen are stored in pressurized tanks. The hydrogen is subsequently used in the fuel cell to produce electricity and also converted into heat for cooking and space heating by catalytic combustion. A novel method of "transparent insulation" retains passive heat, and as an efficient heat collector supplies about 90 percent of the domestic hot water, while the remaining 10 percent is provided by waste heat from the fuel cell. Such new solar technologies will permit construction of buildings with zero or negligible heating energy demand in the future.

266

266. *At the energy-self-sufficient solar house (SSSH) in Freiburg, Germany, surplus energy from the PV generator in summer is fed into an electrolyzer to decompose water, and the resultant oxygen and hydrogen are stored in pressurized tanks (foreground). The hydrogen is subsequently used in fuel cells to produce electricity and also converted into heat for cooking and space heating by catalytic combustion.*

Solar Energy Transmission in Space

The archaic concept in which Heaven and Earth are brought together in a divine unity of opposites has been given a new dimension by the imaginative proposal to convert solar energy in space for transmission to Earth by using a solar power satellite and the wireless power transmission of energy (WPT). These ideas were conceived by Hertz (1857–1894), and WPT was first demonstrated in 1888. Tesla (1865–1943) attempted in vain to demonstrate a global WPT with a tall tower on Long Island, New York, in 1908. Research on WPT could not be pursued until microwave generators were developed in the 1950s; these are now used in more than 250 million microwave ovens worldwide.

It was only when microwave rectifiers capable of converting microwaves with high efficiency were developed in the 1960s that WPT was successfully demonstrated in 1964 with the flight of a small microwave-powered helicopter. A complete WPT system was demonstrated in 1975 at the NASA Deep Space Antenna Facility at Goldstone, California, where a microwave beam at a frequency of 2.45 gigahertz transmitted 30 kilowatts over a distance of 10 miles (1.5 kilometers) to a flat receiving antenna on which microwave rectifiers were mounted. The microwave beam was converted directly into electricity with an average efficiency of 82 percent.

As part of the Stationary High-Altitude Relay Platform (SHARP) program, the Canadian Department of Communications demonstrated in 1987 that an aircraft could be maintained at an altitude indefinitely when powered by a controlled microwave beam. In Japan's WPT development program, several successful test projects were conducted, including WPT to a small aircraft (1992), WPT from a rocket to a satellite launched from the rocket (1993), and WPT to a dirigible (1996). WPT in the microwave and laser sections of the electromagnetic spectrum was investigated as part of the Space Power System (SPS) Concept Development and Evaluation Program by NASA and the U.S. Department of Energy (1975–1980). The positive results of this program, which included technical, economic, and sociological assessments of power from space, were noted by other organizations such as the Institute of Space and Astronautical Science in Japan, the European Space Agency, and Eurospace, as well as institutes and universities in Canada, China, India, Russia, and Ukraine.

An intermediate step in the use of WPT is the power relay satellite (PRS), designed to reflect a microwave beam sent over intercontinental distances from a selected beam transmitter site to a receiver at a desired location. This concept was investigated by the European Community in Brussels in 1991–92. Placed in a geosynchronous orbit position concentric with

---

267. Originally intended for exclusive use in spacecraft, PV systems have since been brought down to Earth for the benefit of humanity as a whole. This cooperative spirit is epitomized by the American space shuttle

Atlantis docking with the Kristall module of the Russian Mir space station, as shown in this technical rendition produced by John Frassanito and Associates for NASA, Washington, D.C.

268. An American sun symbol.

269

the equator, a PRS system provides a transmission link between a major renewable energy site and a distant power grid to meet electrical power requirements — transmitting power, for example, from Australia to India, or from South America to Europe. In this system, electricity is fed to microwave generators incorporated into a phased-array antenna that transmits the microwaves in a controlled beam to the microwave reflector in the PRS. The reflector redirects the beam to the desired location on Earth, where it is converted to electricity and fed to the power grid. The receiving antenna may be located on land or on a floating platform at sea.

The orbiting satellite can also carry a large PV or solar thermal converter and beam the energy produced to a receiver on the ground. The placement in space would result in much higher efficiency than any location on the surface of the Earth. NASA conducted a far-reaching examination of the technologies, system concepts, and terrestrial markets that could be served by space solar power systems, including SPS (1995–96). Working within the context of the global 21st-century energy market, and analyzing about 30 SPS concepts, very promising results emerged based on preliminary market and economic analyses. An outline was provided on how the SPS concepts and technologies might be pursued in the future. As a result, there has been significant progress in applicable technologies, leading to improved system economics and important reductions in environmental impact compared with other contemporary energy sources.

The SPS holds significant promise as a unique solution to preserving the Earth's environment — its most significant advantage being the potential for continuous generation of electricity from satellites in the Earth's orbit within the framework of global communica-

269. In the Space Power System (SPS) concept, PV systems convert solar energy directly into electricity and feed it to microwave generators forming part of a planar, phased-array transmitting antenna. The antenna then directs a beam of very low power density to one or more receiving stations on Earth.

270

tions. It is increasingly likely that in the 21st century SPS could be a critical component of the global energy supply infrastructure with long-term benefits for the world's population. The United Nations and its agencies have developed the legal and regulatory framework, and have the political legitimacy to make SPS development and implementation a truly international effort. The challenge today is not only to arrive at an unbiased assessment of viable options that can meet energy requirements at various stages of human development, but also to recognize that there may be only a limited time left — measured in a few decades — to open up the space frontier so that the benefits of solar energy converted in space for transmission to Earth can be realized. The enormous costs of NASA's space programs would be well justified if some of the funds were diverted to energy-oriented systems of the kind represented by solar energy transmission in space.

271

270. *Land-based receiving antennas consist of mass-produced microwave rectifying elements mounted on an open grid structure. They can be up to 70 percent transparent to sunlight, which permits land underneath the antenna to be utilized for other purposes.*

271. *A Japanese sun symbol.*

Culture of Heliotechnology

A paradigm attributed to the Sumerian priestess Enkheduana, one of history's earliest known authors (c. 2300 B.C.), envisages "paradise as the place where the sun rises" — a concept whose origins must go back some 4 million years to when Mâui-akamai's earliest ancestors learned to walk on two feet and looked with awe at the sun that illuminates the Earth with its life-giving light and warmth. Indeed, the enormous output of solar energy is mind-boggling: if the sun were enveloped in a shell of ice 40 feet (12 meters) thick, it would thaw its way out in less than one minute — and each year the amount of solar energy falling on the Earth is ten times that of all fossil fuel reserves and uranium combined. So the efficient and economical use of solar energy derived from the biomass, from thermal, geothermal, and PV sources, from wind and ocean power, from small hydroelectric facilities, and from hydrogen energy (without fusion) could easily satisfy the world's energy requirements.

A new chapter in Mâui-akamai's Stone Age quest to capture solar energy had opened in the 13th century with the accidental discovery of "lodestone" and its electromagnetic properties, which led to the development of electricity by various means, including the photovoltaic effect of sunlight. It consecrated Mâui-akamai's attempts to lasso the sun on the great volcano Haleakalâ, the "House of the Sun," which, in a modern context, is represented by mountains such as Mont Soleil in the Swiss Alps, where a 500-kilowatt photovoltaic system has been installed, as well as other mountains, such as California's Altamont Pass, the Tarifa ranges in Spain, the Muppandal mountain valleys in India, and similar windswept locations, where thousands of wind turbines are producing electricity. As a result, Mâui's mother, Hina-a-keahi, now washes and dries her son's *kapa* in a washing machine and preserves food in a refrigerator while she watches TV and hears the latest news on the radio. The quality of her life has been enhanced and she inspires others to imitate her lifestyle, not only in the developing countries where innumerable individ-

274

272. *"The sun is the greatest of all visible, earthly fires," stated Zoroaster (c. 628–551 B.C.). In this 1st-century B.C. environmentally friendly Changxin Palace lantern, the rotating deck is adjustable to provide light in different directions, and the smoke is absorbed by water stored in the courtesan's body as it passes through her arm. It was discovered in 1969 in a Han tomb in Hebei Province, China.*

273. *Indian sun symbol used in festivals and cultural events.*

274. *Early Bronze Age rock drawings from Ekenberg, Sweden, depicting solar energy as concentric circles, horses, and boats.*

275

ual and community PV and thermal operating systems have been installed, but also in the industrialized countries, such as the United States, where the Million Solar Roofs campaign has been launched, and Germany, where the BMFT program installs small grid-connected solar arrays on residential and commercial buildings.

Originally intended for use exclusively in spacecraft, photovoltaic systems have not only been brought down to Earth with the help of Mâui-akamai's "scientific lasso," but they have been tied to cultural roots, represented by the old wiliwili tree near his grandmother's home. The traditional relationship of the sun with trees is acknowledged universally and identified with the protection of the environment. In India's arid regions of Rajasthan, the life-sustaining protection of the environment is an article of faith with the white-robed Bishnoi community, the devotees of the sun god Vishnu. They are the ones who inaugurated the *chipko*, or "tree-hugging" movement some 250 years ago, as their village folks valiantly resisted the felling of trees by the Jaipur maharaja's soldiers, who brutally hacked to death 363 Bishnois, along with their brave woman leader, Amrita Devi. She thereby became a legend, like Daphne changing into a laurel, the sacred shrub of the Greek sun god, Apollo, depicted in the beautiful sculpture at the Borghese Gallery in Rome.

The social framework within which a Cacique chief in Panama explains why rainwater (instead of the locally unavailable distilled water) must be used in the battery of his hut's solar panel is simple: "Its power comes from the sun god and therefore the water used for the system must also come from the sky." The Amerindians in western Mexico clean and whitewash their huts not so much to enhance the solar light but to pay homage to the sun god blessing their homes. A Navajo native in the United States echoes the tradition of an Australian aborigine at the other end of the world in his belief that the newly installed solar light in his home helps him to protect his children against sickness. Peasants in southern Asia look upon solar energy projects with

275. A child wearing a jacket with traditional sun motifs poses in front of one of the 288 PV panels of the 56-watt Kankoi Solar Station in Bunair Valley, Pakistan.

276. Chinese children at the Mancheng County Solar Energy School watching the full eclipse of the sun on a PV-run television on March 9, 1997.

277

280

278

279

281

277-286. Solar energy has opened the doors of many schools for children who had been deprived of education and life's future prospects. The pictures show: (277) Indonesian children studying in the school library at Bulak Baru, a village lit by wind turbines. (278) Indian children playing with water from a solar pump. (279) Children of different nationalities in Helsinki using a solar lantern to study together. (280) Japanese children playing with a paper sunflower at the Fukumine kindergarten on the island of Miyako. The school's electricity is entirely supplied by its PV system. (281) An Australian teacher with her students at the Ipolera Aborigine Community school, fitted with a PV system. (282) Finnish children playing in the solar village of

282

285

283

284

286

Kerava near Helsinki. (283) Japanese children at lunch beneath some of the 80 solar tube-lights installed in the Fukumine kindergarten. (284) A farmer in Bunair Valley, Pakistan, telling his son how to sow seeds, with the 280 PV panels of the Kankoi Solar Station in the background. (285) Aborigine children with their solar energy tutor in Alice Springs, Australia. (286) Living in the Rajasthani village of Kadampura with no electricity, these Indian girls never went to school until the newly installed PV lights enabled them to attend night classes.

287

287. *A Mexican woman in the village of San Juan Cancuc*
— near the historic city of San Cristobal de las Casas in
the state of Chiapas— on the way to her village carrying
a PV panel on her back and a child in her arms.

respect and reverence when they are identified with their natural deities — Sûrya, the sun; Vayu, the wind; Agni, the fire; Varuna, the deity of seas and rivers; and the wind god Indra. A cultural link with the renewable energy of the sun is to be found in Europe's fascination with the windmill, as seen in the masterpieces of the Renaissance artists, in Rembrandt's sketches, in the impressionist works of Van Gogh, in Mondrian's paintings, and also in literature such as Cervantes's *Don Quixote*.

Unfortunately, the Polynesian deity has yet to see the advent of the Golden or Green Age of solar energy — the final stage in the Stone-Bronze-Iron-Age sequence. He is deprived of the happy ending to his enchanting saga, primarily as a consequence of the world's exponentially growing population. Mâui-aka-mai would not have found himself in this situation had he taken into account the equation governing population growth and energy —

that increase in population requires corresponding efforts to produce more energy, and that society's economic and social benefits are proportional to energy surpluses, just as energy deficits create poverty. He did not heed the theory on population control expounded about 150 years ago by Thomas Robert Malthus, who rightly stated in his *Essay on the Principle of Population* that "as the rate of increase in human population is in a geometrical ratio, while that of increase in human food supply is only in arithmetical ratio, the result must be misery and death for the poor, unless population growth is checked."

The validity of Malthus's hypothesis is now apparent, as the energy produced worldwide is far less than the energy consumed by the planet's ever growing population, which has increased from 1.5 billion in 1900 to about 5.4 billion today. At this rate the world's population is expected to top 6.25 billion in the year

288

289

288. *Farmers in the Indian village of Kalyanpur in Uttar Pradesh have been able to cook after sunset since a 100-kilowatt PV power station started operating in 1993. It provides the 400 families with 450 domestic lamps, 65 streetlights, and 15 water pumps.*

289. *In one of the remote regions of Inner Mongolia, a herdsman's wife poses with her child in front of a 200-watt PV panel, which is used alternately for three lights, a television set, a small washing machine, and a refrigerator.*

2000 before reaching about 9 billion in 2025. Therein also lies the reason why the poor have more children, for when these illiterate people have no access to usable energy they seek survival through the traditional escape route of greater human muscle power by increasing the size of their families. And Mâui-akamai wonders why nobody has had the courage to devise the means to measure the appalling scale of this energy deficit that is throwing millions of people out of work, and to calculate the unpaid and unaccounted labor of approximately one half the world's population — women.

Moreover, wealth and energy use being two sides of the same coin, the unequal distribution of energy has created unacceptable income disparities between the richest 20 percent and the poorest 20 percent of the world's people. The gap has more than doubled in the last 30 years, going from a ratio of 30-to-1 to 60-to-1. Between 1960

290. Water is the lifeblood of the 70 percent rural population in Africa for whom firewood, crop residues, and animal dung are the primary sources of biomass energy. The picture shows a farm irrigated by a solar water pump in the Mande Province of Mali.

and 1993, the per capita income gap between industrialized and developing countries almost tripled. While 15 countries have seen a surge in energy and economic growth over the past three decades, 1.6 billion people living in more than 100 countries are worse off today than they were 15 years ago — and 2.4 billion people are denied such elementary necessities of life as electricity. The information revolution will inevitably make people more conscious of this appalling gulf between the haves and the have-nots — a gap with disastrous consequences unless a more equitable energy utilization system is quickly put into place.

Another dangerous consequence of the energy deficit caused by population explosion is the avalanche of desperate poor from rural areas who are pouring into energy-starved mega-cities, while the urban unemployed in developing countries are migrating in increasing num-

292. *Living on Uros, "the floating island in the sun," on Lake Titicaca, a Peruvian couple use a PV panel installed on one of the huts, while the sun's thermal energy dries fish on another roof.*

bers to industrialized nations. As a result, there has been a five-fold increase in the number of citydwellers between 1950 and 1990, and it is estimated that more than half the world's population will be living in cities by the year 2000 if the current daily migration of 150,000 people into cities is not halted. United Nations projections of population growth show that by 2025 Egypt's total population will have grown from 51 to 94 million; Nigeria's from 105 to 301 million; Mexico's from 85 to 150 million; and India's from 950 million to nearly 1.5 billion — on a par with China's projected total. The writing on the wall is clear: without more vigorous family planning efforts, worldwide population might increase to 14 billion by the end of the next century, with nearly nine out of ten people living in the developing countries with proportional increases in poverty and misery.

The grinding poverty in which one third of the world's population is surviving on merely one dollar a day — as reported by the World Bank — is fraught with grave socioeconomic and political consequences. The misery and desperation due to lack of energy sources for such exploding numbers of people is exploited by ruthless dictators and religious fundamentalists, while immigration pressures fuel neo-Nazi and fascist movements. The gruesome genocide in Rwanda, in which tens of thousands of innocent men, women, and children were butchered, is not just due to tribal animosity — Hutus and Tutsis have intermarried and lived together for centuries — but was primarily the result of the inability of the life-support energy system to sustain this tiny, densely populated country.

The nonconventional energy of the sun is the only renewable source that can provide the elementary amenities of life and create the tolerable living conditions that may persuade more than 2 billion poor people worldwide not to uproot themselves from their natural habitat and chase an elusive mirage of urban prosperity. Towards this end, small and dispersed solar energy projects in rural areas have a cardinal role to play in stabilizing population growth and halting the headlong rush to the cities. By virtue of their local, decentralized, and participatory nature, sustainable solar energy projects, installed with minimal credit, are inherently "democratic." They tend to create new cooperative structures that counter the concentration of power in a few authoritarian hands controlling gigantic projects. They are the antithesis of the alienating division between suppliers and consumers, between the technical expertise of administrators and the traditional wisdom of common people, between culture and science itself — which is basically a modern, energy-related phenomenon that started with the Industrial Revolution.

The traditional methods of using solar energy inspire people to take pride in their cultural identity and encourage individuals to observe the norms of community living in an interdependent world in which personal freedom is inseparable from collective obligations towards human and animal life. This is reflected in the Japanese Shinto concept of *kami* — representing the deities of both Heaven and Earth, who appear in ancient records as human beings and, at the same time, as birds, beasts, plants, seas, mountains, stones, trees, water, and so forth. They are integral parts of the ecological balance of Nature's "worldwide web," in which all things are interrelated. In the Persian Pahlavi concept of genesis, humankind has the lowest status in Nature's order of precedence — the sun (sky), water, the Earth, plants, animals (cattle), and humankind — thereby reiterating our primary obligations towards sustainable sources of solar energy.

This is also the moral of Mâui-akamai's simple tale, which calls for the use of renewable solar energy for the sustenance of life and protection of the environment — menaced as they are by the population explosion, abject poverty, urban over-

292. With little government support, some private American companies, such as Sundance Solar Designs, have taken the initiative to install photovoltaic systems in homes like this one owned by a Navajo couple in Colorado. The company has now extended its operations to the Caribbean, reinforcing the efforts of Enersol Associates and a Dominican organization, Asociación para el Desarollo de Energía Solar, whose "revolving credit fund" has helped to equip more than 2,000 homes in the Dominican Republic with solar power since 1984.

292

293

crowding, moral degradation, and religious and ethnic extremism, which are undermining democratic institutions. It teaches us that only through the use of traditional heliotechnology can we recreate the rhythm of nature that flows through the world's biosphere, or zone of life, which excessive exploitation of fossil fuels has been destroying since the Industrial Revolution. Learning from the past can help us to plan

for the future — by acting in the present while we still have the time and opportunity to forge ahead with carefully conceived and adequately funded solar energy projects. For there is no other way to ensure that our "global village" will remain a habitable home for humanity except by preserving the ecological equilibrium of our planet through use of the renewable energy of the sun that sustains all forms of life on Earth.

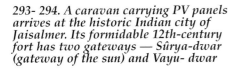

293- 294. A caravan carrying PV panels arrives at the historic Indian city of Jaisalmer. Its formidable 12th-century fort has two gateways — Sûrya-dwar (gateway of the sun) and Vayu- dwar

(gateway of the wind). An annual festival to celebrate the shakti (energy) of the sun and the wind is held at Sûrya- dwar, where the image of the sun is worshipped.

294

Although the terms "energy" and "power" are often used interchangeably in everyday conversation, each term has a distinct meaning. Energy is defined as the capacity for doing work, such as lifting an object a certain distance or heating a substance. Power, on the other hand, is the amount of work performed per unit of time. The British thermal unit, or Btu, and the kilowatt-hour are energy units commonly used to describe the quantities of energy produced or consumed in ordinary situations, such as the energy consumed in heating, cooling, or lighting a home or an office. The unit of power most commonly used for such situations is the kilowatt.

One Btu is the amount of energy required to warm 1 pound of water 1 degree Fahrenheit. Since the precise definition of a kilowatt-hour is difficult to visualize, the kilowatt-hour can be regarded as the amount of energy consumed in lighting ten 100-watt bulbs for a period of one hour. A kilowatt, on the other hand, is the rate at which energy must be used to light ten 100-kilowatt bulbs.

The terms "megawatt" (1 million watts) and "gigawatt" (1 billion watts) are commonly used to describe the generating capacity of power plants. A large nuclear or coal-powered plant, for example, might have a capacity of 1,000 megawatts (1 gigawatt) and would produce 8.76 billion kilowatt-hours of energy in one year if it operated continuously at full power during that time.

In discussions of the annual energy consumption of an entire nation or the amount of energy contained in a major energy resource, the quad — short for 1 quadrillion (10^{15}) Btu — is often used. One quad is a massive amount of energy, equivalent to about 293 billion kilowatt-hours. In 1990, for example, the total amount of energy used worldwide (excluding energy from fuelwood, charcoal, and alcohol) was about 321 quads, and the world's largest energy consumer, the United States, used about 79 quads.

Note by Richard Golob and Eric Brus in *The Almanac of Renewable Energy*

Acknowledgments

WORKS OF ART CREDITS

ABC— Mondrian Estate Holtzman Trust, New York. Haags Gemeentemuseum, the Hague, Holland: 100

Borghese Gallery, Rome, Italy: 14

British-India Office Library, London, England: 25

British Museum, London, England: 15 16

Conservatoire National des Arts et Métiers, Paris, France: 161

Dietrich Varez's woodcuts from the book *Mâui*. Bishop Museum, Honolulu, Hawaii, USA: 6

Elizabeth Schultz Collection, Lawrence, Kansas, USA: 139

G. Pouw's illustrations, Zaandam, Holland: 70

Louvre Museum. Réunion des Musées Nationaux, Paris, France: 13 68 120

Musée des Antiquités Nationales, Saint-Germain-en-Laye. Réunion des Musées Nationaux, Paris, France: 19

National Gallery of Art, Washington D.C., USA: 180

National Museum, Copenhagen, Denmark: 26

National Museum of Korea, Seoul: 119

Observatoire de Paris, France: 162

Rijksmuseum, Amsterdam, Holland: 72 73 75 93 95

Riksantikvarieambetet, Stockholm, Sweden: 274

Sotheby's, New York, USA: 174

Stanley Collection of African Art, University of Iowa Museum of Art, Iowa City, USA: 11

Sungjon University Museum, Republic of Korea: 156

Uffizi Gallery, Florence, Italy: 122

Van Gogh Museum, Amsterdam, Holland : 42

PHOTOGRAPH CREDITS
Individuals

Maurice Aeschimann, Geneva, Switzerland: 2 34

Karidin Akmataliyev, Frunze, Kirgizstan: 12

William H. Avery, The Johns Hopkins University, Laurel, USA: 134 138

H. Baranger—Phototheque EDF, La Richardais, France: 133

Linda Bercusson, Christchurch, New Zealand: 89

Michel Brigaud—Phototheque EDF, La Richardais, France: 132

Rémy Delacloche—Fondation Energies pour le Monde, Paris, France: 1 5 113 221 290

Steven Andre Dibner, San Francisco, USA: 291

Nejat Diyarbrkirli, Istanbul, Turkey: 38

Ben Gibson / Katz —Cosmos, Paris, France: 31

Peter E. Glaser, Cambridge, MA, USA: 269 270

A. Goetzberger, Freiburg, Germany: 239 266

Rafn Hafnfjörd, Reykjavik, Iceland: 146 147 148 149

Flemming Hagensen, Risø National Laboratory, Roskilde, Denmark: 86 87

Mike Jackson, Border Wind Limited, England: 88

Toshio Jo, Asahi Evening News, Tokyo, Japan: 224 225

Eiji Kitada, Koganei, Tokyo, Japan: 187

François Kronenberger, Strasbourg, France: 259

Larry Lee—COSMOS, Paris, France: 69

Even Mehlum, SINTEF, Oslo, Norway: 128

Pavan Mehta, New Delhi, India: 205

Peter Menzel—COSMOS, Paris, France: 30 171

Khojeste Mistree—Zoroastrian Studies, Bombay, India: 154

S. O'Dowd, Christchurch, New Zealand: 212

Claude Pauquet—Phototheque EDF, La Richardais, France: 131

Jan L. Perkowski, New York, USA: 17

F. Pharabod, Paris, France: 169

Roger H. Ressmeyer— Starlight/COSMOS, Paris, France: 81 140 152 172 254

Riesjard Schropp—NOVEM, (European Community THERMIE Programme), Utrecht, Holland: 59 117 175 176 177 178 179 195 196 203 207 208 210 215 223 226 227 228 240 241 243

Madanjeet Singh, Paris, France: 3 4 7 9 14 15 16 18 20 21 22 23 24 32 33 35 36 39 40 41 45 46 47 48 49 50 55 56 58 64 77 78 79 80 82 83 90 91 92 94 96 102 109 110 111 114 115 116 118 136 141 155 158 159 160 163 165 166 167 168 170 182 184 185 186 188 189 190 191 194 204 206 211 222 230 231 232 233 234 235 236 237 238 246 247 248 249 250 251 253 257 268 271 272 273 275 276 277 278 279 280 281 282 283 284 285 286 288 289 292 293 294

Duby Tal, Tel Aviv, Israel: 164

Hua Tao, Nanjing, China: 29

Vladimir Terebenin, St. Petersburg, Russia: 27

Sheila Terry—Science Photo Library, London, England: 28

J.L.J. Tersteeg, Rotterdam, Holland: 71 76 97 98 99

Akira Ueda, OM Solar Association, Hamamatsu City, Japan: 197 198 199

Francesco Venturi—KEA Publishing, London: 10

Jacques Verroust, Paris, France: 8 9

Duncan Willetts—COSMOS, Paris, France: 183

Sarab Zavaleta, New York, USA: 181

PHOTOGRAPH CREDITS
Institutions

Adam Editions, Athens, Greece: 65

Comité d'Action pour le Solaire, Paris, France: 202

Commonwealth Department of Energy, Canberra, Australia: 244

Daimler-Benz AG, Sttutgart, Germany: 252 255 256

Energy Conversion Devices, Inc., Troy, Mich., USA: 209

Energy, Resources, and Technology Division, State of Hawaii, Honolulu: 37 52 53 54 60 101 112 135 137 153 229 245 263

Fjellanger Wilderøe AS, Oslo, Norway: 121

Hans Grohe, Schiltach, Germany: 200

Honda Motor Company, Tokyo, Japan: 260

Hydro Québec, Montréal, Canada: 104

Itaipu Binacional, Rio de Janeiro, Brazil: 103 105 106

Jet Propulsion Laboratory, California Institute of Technology, Pasadena, USA: 173

Kvaerner AS, Oslo, Norway: 123 124

NASA, Washington, D.C., USA: 267

NEDO, Tokyo, Japan: 129 130 143 144 145 150

Neste Advanced Power Systems, Helsinki, Finland: 85 193

Nordtank Energy Group, Denmark: 84

Norwave AS, Oslo, Norway: 125 126 127

Rehia Photo Library, Ruawae Northland, New Zealand: 258

Risø National Laboratory, Roskilde, Denmark: 66 67 74

Shinto Shrine, Ise, Japan: 43 44 157

Siemens Solar Industries, Camarillo, Calif., USA: 213 242 264

Swiss PV-Tour, Geneva, Switzerland: 201

UKAEA, London, England: 107 108

Union Fenosa, Madrid, Spain: 63

United Solar System Corp., San Diego, USA: 192 262 287

Weizmann Institute of Science, Rehovot, Israel: 265

ZREU—(European Community THERMIE Programme), Regensburg, Germany: 51 57 61 62

Bibliography

Reports of the UNESCO High Level Expert Meetings (1993–1996) held in Zimbabwe, China, Israel, Malaysia, Pakistan, Oman, Malta, Costa Rica, Russia, and Japan to formulate the plan of action for the World Solar Programme 1996–2005.

A Golden Thread (1980). Ken Butti and John Perlin. Cheshire Books, Palo Alto, CA, U.S.A.

Solar Racing Cars (1994). J.W.B. Storrey, A.E.T. Schinckel, and C.R. Kyle. Australian Government Publishing Service, Canberra, Australia.

Consumer Guide to Solar Energy (1995). Scott Sklar and Kenneth Sheinkopf. Bonus Books, Inc., Chicago, U.S.A.

The Almanac of Renewable Energy (1993). Richard Golob and Eric Brus. Henry Holt and Company, New York, U.S.A.

Global Change of Planet Earth (1994). Organization for Economic Co-operation and Development, Paris, France.

Policy Implications of Greenhouse Warming (1991). National Academy of Sciences. National Academy Press, Washington, DC, U.S.A.

Power Surge (1994). Christopher Flavin and Nicolas Lenssen. W.W. Norton and Company, New York, U.S.A.

The Dutch Windmill (1963). Frederick Stokhuyzen. Universe Books, Inc., New York, U.S.A.

Mâui (1991). Dietrich Varez. Bishop Museum Press, Honolulu, Hawaii, U.S.A.

L'énergie solaire en France (1993). Alexandre Herléa. Édition du cths, Paris, France.

Renewable Energy and Energy Efficiency in Latin America and the Caribbean (1994). U.S. Department of Energy, Washington, DC, U.S.A.

Renewable Energy Technologies in China (1994). Government of China Publications, Chongquin Changjiang Printing Company, Beijing, China.

A Solar Manifesto (1994). Hermann Scheer. James and James Publishers, London, England.

Passive Solar Energy (1994). Bruce Anderson and Malcolm Wells. Brick House Publishing Company, Amherst, NH, U.S.A.

Photovoltaic Fundamentals (1991). Gary Cook, Lynn Billman, and Rick Adcock. U.S. Department of Energy, Washington, DC, U.S.A.

Renewable Energy: Sources for Fuels and Electricity (1993). Thomas B. Johansson, Henry Kelly, Amulya K.N. Reddy, and Robert H. Williams. Island Press, Washington DC, U.S.A.

Geothermal Resources (1987). David N. Anderson and John W. Lund. California Geothermal Resources Council, U.S.A.

Solar Hydrogen: Moving Beyond Fossil Fuels (1989). Joan A. Ogden and Robert H. Williams. World Sources Institute, Washington, DC, U.S.A.

Energy and the Environment into the 1990s (1990). A.A.M. Sayigh. Pergamon Press, Oxford, England.

Energy for Planet Earth (1991). Scientific American Readings. W. H. Freeman and Company, New York, U.S.A.

Wind Energy for a Growing World (1990). American Wind Energy Association, Washington, DC, U.S.A.

Global Energy Perspectives 2000–2020 (1989). World Energy Council, London, England.

World Oil: Coping with the Dangers of Success (1985). Christopher Flavin. Worldwatch Institute, Washington, DC, U.S.A.

Fueling One Billion: An Insider's Story of Chinese Energy Policy (1993). Yingzhong Lu. Washington Institute, Washington, DC, U.S.A.

The New Oil Crisis and Fuel Economy Technologies (1988). Deborah Lynn Bleviss. Quorum Books, Westport, CT, U.S.A.

What Is a Fuel Cell? (1992). Philip H. Abelson. Fuel Cell Commercialization Group, Washington, DC, U.S.A.

World Progress in Wave Enegy (1988). S.H. Salter. Department of Mechanical Engineering, University of Edinburgh, Scotland.

Advanced Energy Systems and Technologies Research in Finland (1994–1995). NEMO 2 Annual Report, Helsinki, Finland.

Sun in Action (1995). A Strategic Plan for Action in Europe, Report of the Commission of the European Communities. Director General for Energy, Brussels, Belgium.

Ocean Thermal Energy Conversion (1994). Luis A. Vega. Pacific International Center for High Technology Research, Honolulu, Hawaii, U.S.A.

Solar Energy Utilization Technology (1994). New Energy and Industrial Technology Development Organization (NEDO), Tokyo, Japan.

Solar Energy in Israel (1991). David Faiman. Ben-Gurion National Solar Center, Sede Boqer Campus, Israel.

Index

Printed and bound by
Amilcare Pizzi arti grafiche S.p.A.
Cinisello Balsamo (Milan)
Finished printing in January 1998